Surprising Geometric Solutions

A Projective Geometry Approach

Problem Solving in Mathematics and Beyond

Print ISSN: 2591-7234
Online ISSN: 2591-7242

There are countless applications that would be considered problem solving in mathematics and beyond. One could even argue that most of mathematics in one way or another involves solving problems. However, this series is intended to be of interest to the general audience with the sole purpose of demonstrating the power and beauty of mathematics through clever problem-solving experiences.

Each of the books will be aimed at the general audience, which implies that the writing level will be such that it will not be engulfed in technical language — rather the language will be simple everyday language so that the focus can remain on the content and not be distracted by unnecessarily sophiscated language. Again, the primary purpose of this series is to approach the topic of mathematics problem-solving in a most appealing and attractive way in order to win more of the general public to appreciate this most important subject rather than to fear it. At the same time we expect that professionals in the scientific community will also find these books attractive, as they will provide many entertaining surprises for the unsuspecting reader.

Published

Vol. 46 *Surprising Geometric Solutions: A Projective Geometry Approach*
by Alfred S Posamentier and David Stuhlpfarrer

Vol. 45 *How to Generate a Genuine Love of Mathematics*
by Alfred S Posamentier

Vol. 44 *Revolutions in Mathematics*
by Arthur D Kramer and Alfred S Posamentier

Vol. 43 *Mathematical Puzzles and Curiosities*
by Ivo Fagundes David de Oliveira, Tanya Khovanova and Yogev Shpilman

Vol. 42 *Learning Geometry Through Problem Solving*
by Boris Pritsker

For the complete list of volumes in this series, please visit www.worldscientific.com/series/psmb

Problem Solving in Mathematics and Beyond
Volume 46

Surprising Geometric Solutions

A Projective Geometry Approach

Alfred S. Posamentier
The City University of New York, USA

David Stuhlpfarrer
University College of Teacher Education Styria, Austria

World Scientific

NEW JERSEY • LONDON • SINGAPORE • BEIJING • SHANGHAI • HONG KONG • TAIPEI • CHENNAI • TOKYO

Published by

World Scientific Publishing Co. Pte. Ltd.
5 Toh Tuck Link, Singapore 596224
USA office: 27 Warren Street, Suite 401-402, Hackensack, NJ 07601
UK office: 57 Shelton Street, Covent Garden, London WC2H 9HE

Library of Congress Cataloging-in-Publication Data
Names: Posamentier, Alfred S. author | Stuhlpfarrer, David author
Title: Surprising geometric solutions : a projective geometry approach /
Alfred S. Posamentier, The City University of New York, USA,
David Stuhlpfarrer, University College of Teacher Education Styria, Austria.
Description: New Jersey : World Scientific, [2026] | Series: Problem solving in
mathematics and beyond, 2591-7234 ; volume 46 | Includes index.
Identifiers: LCCN 2026016652 | ISBN 9789819831838 hardcover |
ISBN 9789819831845 ebook | ISBN 9789819831852 ebook other
Subjects: LCSH: Geometry, Projective
Classification: LCC QA471 .P827 2026
LC record available at https://lccn.loc.gov/2026016652

British Library Cataloguing-in-Publication Data
A catalogue record for this book is available from the British Library.

For any available supplementary material, please visit
https://www.worldscientific.com/worldscibooks/10.1142/14839#t=suppl

Desk Editors: Kannan Krishnan/Gabriel Rawlinson

Typeset by Stallion Press
Email: enquiries@stallionpress.com

About the Authors

Alfred S. Posamentier was most recently a Distinguished Lecturer at New York City College of Technology of the City University of New York. Prior to that he was Executive Director for Internationalization and Funded Programs at Long Island University, New York. This was preceded by 5 years as Dean of the School of Education and Professor of Mathematics Education at Mercy University, New York. Before that he was for 40 years at the City College of the City University of New York, at which he is now Professor Emeritus of Mathematics Education and Dean Emeritus of the School of Education. He is the author and co-author of more than 90 mathematics books for teachers, secondary and elementary school students, as well as the general readership. Dr. Posamentier is also a frequent commentator in newspapers and journals on topics related to education and mathematics.

After completing his B.A. degree in mathematics at Hunter College of the City University of New York, he took a position as a teacher of mathematics at Theodore Roosevelt High School (Bronx, New York), where he focused his attention on improving the students' problem-solving skills and at the same time enriching their instruction far beyond what the traditional textbooks offered. During his six-year tenure there, he also developed the school's first mathematics teams (both at the junior and senior level). He is still involved in working with mathematics teachers and supervisors, nationally and internationally, to help them maximize their effectiveness.

Immediately upon joining the faculty of the City College of New York in 1970 (after having received his master's degree there in 1966), he began to develop in-service courses for secondary school mathematics teachers, including such special areas as recreational mathematics and problem solving in mathematics. As Dean of the City College School of Education for 10 years, his scope of interest covered the full gamut of educational issues. During his tenure as Dean, he took the School of Education from the bottom of the New York State rankings to the top with a perfect NCATE accreditation assessment in 2009. He also raised more than $12,000,000 from the private sector for innovative educational programs. Posamentier repeated this successful transition at Mercy University, where he enabled it to become the only college to have received both NCATE and TEAC accreditation simultaneously.

In 1973, Dr. Posamentier received his Ph.D. from Fordham University (New York) in mathematics education, and he has since extended his reputation in mathematics education to Europe. He has been visiting professor at several European universities in Austria, England, Germany, the Czech Republic, Turkey and Poland. In 1990, he was Fulbright Professor at the University of Vienna.

In 1989 he was awarded an Honorary Fellow position at the South Bank University (London, England). In recognition of his outstanding teaching, the City College Alumni Association named him Educator of the Year in 1994, and in 2009. New York City had the day, May 1, 1994, named in his honor by the President of the New York City Council. In 1994, he was also awarded the *Grosse Ehrenzeichen für Verdienste um die Republik Österreich* (Grand Medal of Honor from the Republic of Austria), and in 1999, upon approval of Parliament, the President of the Republic of Austria awarded him the title of University Professor of Austria. In 2003, he was awarded the title of *Ehrenbürgerschaft* (Honorary Fellow) of the Vienna University of Technology, and in 2004 he was awarded the *Österreichisches Ehrenkreuz für Wissenschaft & Kunst 1. Klasse* (Austrian Cross of Honor for Arts and

Science, First Class) from the President of the Republic of Austria. In 2005, he was inducted into the Hunter College Alumni Hall of Fame, and in 2006 he was awarded the prestigious Townsend Harris Medal by the City College Alumni Association. He was inducted into the New York State Mathematics Educator's Hall of Fame in 2009, and in 2010 he was awarded the coveted Christian-Peter-Beuth Prize from the Technische Fachhochschule — Berlin. In 2017, Posamentier was awarded *summa cum laude nemine discrepante* by the Fundacion Sebastian, A.C., Mexico City, Mexico.

He has taken on numerous important leadership positions in mathematics education locally. He was a member of the New York State Education Commissioner's Blue Ribbon Panel on the Math-A Regents Exams, and the Commissioner's Mathematics Standards Committee, which redefined the Mathematics Standards for New York State, and he also served on the New York City Schools' Chancellor's Math Advisory Panel.

Dr. Posamentier is still a leading commentator on educational issues and continues his long-time passion of seeking ways to make mathematics interesting to teachers, students and the general public — as can be seen from some of his more recent books.

For more information and a list of his publications see: https://en.wikipedia.org/wiki/Alfred_S._Posamentier.

David Stuhlpfarrer is an Austrian mathematician and mathematics educator with a strong focus on geometry, spatial reasoning, and the didactics of mathematics. His academic and professional career combines long-standing experience in secondary education with extensive involvement in teacher education, educational research, and national quality assurance initiatives.

After completing his secondary education in Graz, he studied Mathematics and Descriptive Geometry at Graz University of Technology and the University of Graz. He graduated with distinction and was awarded the degree *Magister rerum naturalium* (Mag. rer. nat.).

In 2023 he earned his Ph.D. degree with distinction. His doctoral research focused on the relationship between spatial ability and mathematical achievement. In addition to his formal degrees, he completed several advanced professional development programs in higher education teaching and digital didactics.

David Stuhlpfarrer has been active in education since 2008, teaching mathematics and geometry across a broad range of institutions. His professional experience includes long-term appointments in academic secondary schools, vocational and adult education, and higher education. For many years, he taught mathematics, geometric drawing, and descriptive geometry at an academic secondary school in Graz. Alongside this, he worked as a trainer, examiner, and mentor in adult education and in programs preparing learners for a nationally recognized higher education entrance qualification, including teaching assignments in health-care and technical training institutions.

Between 2017 and 2020, he worked as a university assistant at the University of Graz, contributing to research and teaching in mathematics education. He has been appointed as a lecturer at Graz University of Technology, where he taught methodology and didactics courses in descriptive geometry for prospective teachers. Since 2020, he has been a faculty member at the University College of Teacher Education, Styria, where he is involved in both pre-service and in-service teacher education for mathematics at the secondary level.

In parallel with his teaching activities, David Stuhlpfarrer has been extensively involved in educational development and large-scale assessment. He has worked as an item writer for national standardized assessments in mathematics and as a quality observer for competence-oriented evaluations in primary and secondary education. His work in this area contributes to the development, implementation, and evaluation of standards-based mathematics education in Austria.

He is actively engaged in professional networks and academic associations related to geometry and mathematics education. For many years, he has contributed to the Thematic Network Geometry (TNG), including service as coordinator, and has held leadership and advisory roles within the Austrian Association for Geometry (ADG).

He is also involved in the Austrian Society for Subject Didactics (OEGFD), representing the field of geometry.

David Stuhlpfarrer has published widely in the field of mathematics education. His publications include journal articles, practitioner-oriented contributions, conference posters, and an international book. His work addresses topics such as geometry education, spatial reasoning, hands-on and inquiry-based learning, and the integration of digital and physical tools in mathematics instruction, with a strong emphasis on bridging research and classroom practice.

Contents

About the Authors		v
Introduction		xiii
Chapter 1	Overview of Projective Geometry	1
Chapter 2	Geometric Essentials Revisited	7
Chapter 3	Euclid's Fifth Postulate: Various Models of Geometry	35
Chapter 4	Innovative Geometric Techniques Using Projective Geometry	55
Chapter 5	Surprising Solutions to Geometrical Problems	79
Chapter 6	The Cross-Ratio and Its Application in Locating Midpoints	95
Chapter 7	Homogeneous Coordinates: A Concept to Better Understand Infinity	115
Chapter 8	Conic Sections Revisited	135
Index		155

Introduction

Projective geometry includes Euclidean geometry as well as the more modern hyperbolic and elliptic geometries, with the latter two exceeding Euclid's famous fifth postulate. People typically use Euclidean geometry in preparing plans for the construction of a building. They sketch points, straight lines, and circles on a sheet of plane paper. They can use geometrical constructions they have learned at school using an unmarked straightedge and a pair of compasses, for example, to determine the midpoint of a line segment. A pilot or an astronomer, for example, will obviously rely on spherical geometry, since we live on a sphere, which is a form of elliptic geometry, where in some instances other laws of geometry are valid. Projective geometry encompasses many aspects of these various geometries and provides powerful tools that are not well known — even by those who are genuinely interested in geometry and in solving geometrical problems. In this book, major ideas of projective geometry will be introduced and demonstrated by applying them to easily comprehensible geometric problems. The solutions to these problems are not only surprising in the way they can be found, but the results themselves may also be astonishing to the reader. For instance, the concept of figuring out the solution to special geometrical problems by "trying three times" and then using the Steiner circle to get the exact solution out of these random attempts may be particularly surprising. Over the course of solving these problems, the concepts of homogeneous coordinates and the Steiner circle will be introduced and explained on an intelligible level.

The world around us is full of perspective. Pictures of railway tracks seem to converge in the distance. A round coin looks like an ellipse when tilted. Shadows stretch and shrink depending on the angle of the sun rays. These are not optical illusions — they are manifestations of a deeper geometric truth. This truth lies not within the familiar realm of Euclidean geometry, but in the broader, more flexible framework of projective geometry.

At first glance, projective geometry might seem like a specialized corner of mathematics — something distant from everyday experience. In reality, it is the geometry of vision, structure, and transformation. It captures the rules that govern how we perceive depth, how cameras translate 3-D scenes into flat images, how architects draw space, and how artists render perspective. What Euclid describes with an unmarked straightedge and a pair of compasses, projective geometry generalizes with lines, intersections, and vanishing points. It is, in a sense, the geometry of the infinite.

This book invites the reader into the world of projective geometry not only through rigorous constructions, but also through visual intuition and historical insight. We begin in the concrete world of figures and proportions, rooted in the classical Greek tradition. In the second chapter, fundamental geometric ideas — lengths, angles, similarity, proportions, and constructions — are revisited and re-examined with fresh eyes. These essentials serve both as a bridge from familiar concepts and as a foundation for the abstract principles to come.

One of the central themes of projective geometry is duality: the idea that every statement about points has a corresponding statement about lines, and vice versa. Another is invariance: projective transformations may bend and warp figures, but certain relationships — such as the cross-ratio of four collinear points — remain unchanged. These principles form the backbone of a geometry that is elegant, abstract, and surprisingly powerful.

Projective geometry also serves as a bridge between disciplines and between ways of thinking. In mathematics, it is foundational to algebraic geometry, where curves and surfaces are studied in projective space to handle their behavior at infinity. In physics, it appears in the structure of spacetime and light cones. In computer science, it underlies computer vision, graphics, and machine perception. In art and architecture, it governs perspective, proportion, and illusion.

The structure of this book reflects this breadth. It begins with foundational ideas — proportions, angles, figures — and builds up toward projective principles. Along the way, we encounter classical constructions, theorems of Pappus and Desargues, properties of conic sections, and the deep symmetry of harmonic divisions. We introduce tools like the Steiner circle for solving projective transformations, and homogeneous coordinates for algebraic formulations. Each chapter draws connections between intuitive geometry, mathematical structure, and real-world application.

This is not merely a textbook, but an invitation to see geometry from a new perspective — literally. Whether you are a student, a teacher, a mathematician, a physicist, or simply someone with a fascination for patterns and space, this journey through projective geometry offers moments of clarity, beauty, and surprise.

Let us begin — not with distance, but with alignment. Not with measurement, but with connection. Geometry, after all, is not only about what we see — but how we see.

Chapter 1

Overview of Projective Geometry

Projective geometry, though originating in the study of perspective and vision, has evolved into a deep and wide-ranging field that bridges mathematical theory, scientific modeling, and artistic practice. By moving beyond measurements of length and angle and focusing instead on incidence, alignment, and transformation, it provides a richer view of space — one where points at infinity are no longer unreachable abstractions but natural components of a unified geometric universe.

1.1 Mathematical and Geometrical Significance

Projective geometry begins with a simple yet profound departure from classical Euclidean thinking: it removes the special role of parallel lines. As presented in Chapter 3, Euclid's Fifth Postulate — the foundation of classical geometry — is replaced by the principle that any two lines meet, possibly at a point at infinity. This conceptual shift leads to a geometry based not on distance or angle, but on incidence and alignment. Within this framework, the axioms of the projective plane, introduced in Chapter 4, enable a consistent and complete system in which constructions and transformations such as perspectivities and projectivities take center stage.

These transformations — mappings that preserve incidence — form the structural language of projective geometry and allow for elegant classification of configurations on lines, conics, and more.

A particularly striking example of this elegance appears in Chapter 5, where two geometric problem settings demonstrate how autoprojectivities (projective self-maps of a line) can be solved using the Steiner circle. In each case, given partial information about fixed points or images, the Steiner circle provides a purely geometric path to complete the transformation. These examples highlight how seemingly abstract projective mappings become tangible through construction, bringing together precision and intuition in a way that is unique to projective geometry.

Another central invariant is the cross-ratio, described in Chapter 6. Defined for four collinear points, it remains unchanged under projective transformations and serves as a key to defining harmonic sets — configurations where the cross-ratio equals –1. Harmonic points play a fundamental role in many constructions and proofs and form the basis for deeper symmetry considerations in projective geometry.

In order to bridge synthetic and analytic approaches, Chapter 7 introduces homogeneous coordinates. By representing points in the projective plane as triplets of numbers (up to scaling), homogeneous coordinates allow us to translate geometric intuition into linear algebra. Equations of lines, intersections, and transformations become computable, and projective invariants become accessible in both symbolic and numerical contexts.

The unifying power of projective geometry reaches its peak in Chapter 8 with the treatment of conic sections. What appears as different curves in classical geometry — circles, ellipses, parabolas, and hyperbolas — become projectively equivalent, transformed into one another by suitable projectivities. This reveals conics not as a fragmented family, but as a single, richly structured geometric entity. Their properties, including tangents, foci, and poles, emerge naturally from the projective view.

Beyond these foundational contributions, projective geometry remains a central structure in many branches of higher mathematics.

It supports large parts of algebraic geometry, where solutions to polynomial equations are studied in projective space to avoid complications at infinity. In complex geometry, moduli spaces, and birational geometry, projective varieties form the natural setting for deep structural insights. Furthermore, the ideas of duality, intersection theory, and line bundles — essential in modern geometry — have their roots in projective constructions. Even in topology and differential geometry, projective concepts reappear. Projective models describe curved surfaces and the geometry of geodesics; they appear in models of non-Euclidean geometries (like hyperbolic and elliptic geometry), and in projective differential geometry, which studies properties invariant under projective transformations. Projective spaces also arise naturally in mathematical physics, including twistor theory and the study of conserved quantities.

Thus, projective geometry is not only a classical theory with elegant constructions — it is a living language that continues to inform and unify major areas of modern mathematics.

1.2 Scientific and Technological Perspectives

The transition from abstract geometry to real-world applications is especially seamless in the natural sciences and technical fields. In optics, for instance, projective principles govern how three-dimensional scenes are projected onto two-dimensional surfaces — a fact not just useful in art but fundamental in physics and image formation. The rules of perspective projection, where lines meet at vanishing points and parallel lines converge at infinity, are core examples of projective mappings in action.

Modern computer vision builds on these ideas. In stereo imaging, motion tracking, and depth reconstruction, the camera is modeled as a projective device. The cross-ratio plays a direct role in estimating depth from multiple views, while projective transformations describe how an image changes with shifts in perspective. As covered in Chapter 7, homogeneous coordinates make these operations algebraically manageable, providing a common language for rendering, animation, and photogrammetry.

Even in theoretical physics, projective concepts appear. In special relativity, light cones and spacetime diagrams often adopt projective compactifications, where infinities are brought into a finite model. Symmetries and causal relationships in spacetime resemble those in projective geometry, making the mathematical structures surprisingly aligned.

In robotics, coordinate transformations — especially involving projections between sensors, actuators, and environments — are modeled using projective tools. Homogeneous matrices describe movements and camera projections alike. Tasks such as object recognition or motion planning rely on understanding how projective views shift across space.

1.3 Artistic and Architectural Contexts

Projective geometry has always had a strong connection to the arts. Its origin lies in the analysis of linear perspective, which revolutionized painting during the Renaissance. What artists developed intuitively — the convergence of lines at vanishing points, the shrinking of parallel lines, and the positioning of figures in space — now finds its precise foundation in projective theory, particularly in the axioms and perspectivities of Chapter 4.

These principles enabled artists to render space with astonishing realism. But beyond realism, projective constructions also allow experimentation. In works of anamorphic art, images are distorted in a way that only resolves correctly from a particular viewpoint — an exact realization of projective transformation. The theoretical explanation for this phenomenon lies in how projections preserve incidence but distort angle and distance — a core idea in projective geometry.

Artists also made use of harmonic division, a concept from Chapter 6, to create compositions with visual balance and proportion. Without measuring, one could divide a segment into segments with special ratios using straightedge alone. This principle appears in classical architecture and in the design of elements such as columns, windows, and facades.

Conic sections, introduced in Chapter 8, are so central to projective geometry and also manifest in architectural structures. Elliptical arches, parabolic mirrors, and hyperbolic vaults are not only structurally sound but visually striking. Projective geometry provides architects with the tools to design such forms with both aesthetic and functional clarity.

One of the most remarkable artists who explored projective distortions was Maurits Cornelis Escher (1898–1972), a Dutch graphic artist whose work bridges the gap between mathematics and visual art. Escher's images often play with impossible architecture, recursive patterns, and spatial paradoxes — effects that rely deeply on the principles of projective geometry. Though Escher had no formal training in mathematics, he intuitively grasped complex geometric ideas, including symmetry groups, perspective transformations, and tilings, and translated them into some of the most iconic visual puzzles of the 20th century.

Projective geometry offers not just a mathematical system, but a way of seeing. Whether solving abstract problems, interpreting physical phenomena, or designing aesthetic experiences, it provides a common language of structure, transformation, and symmetry. Across its many chapters — from the Euclidean foundation in Chapter 3 to the conic elegance of Chapter 8 — projective geometry reveals itself as a deeply interdisciplinary tool. It reminds us that geometry is not only about what we measure, but about how we perceive, how we construct, and how we imagine space.

Chapter 2

Geometric Essentials Revisited

In the interest of providing the readership all the necessary "equipment" for the proofs in this book, and cognizant of the fact that some readers may have forgotten a few of the basics from the secondary school geometry curriculum, we present this geometry review. First, we will review some of the facts and principles you should recognize from your high school geometry class. These results are presented without proof, as the reader's recollection should make them once again familiar. Naturally, any high school geometry book will furnish further understanding of these basic concepts. Next, we present a number of more sophisticated theorems from basic trigonometry. Some readers may just refer back to this chapter when encountering somewhat unfamiliar concepts in the remaining chapters of the book.

2.1 Congruence of Triangles

Triangles can be proved congruent by showing corresponding parts equal:

- Two sides and their included angle (SAS)
- Two angles and their included side (ASA)
- Two angles and a side not included (AAS)
- Three sides (SSS)
- For right triangles: the hypotenuse and a leg (HL)

2.2 Similarity of Triangles

Triangles can be proved similar by showing any of the following equivalent properties:

- The corresponding angles are equal (AAA).
- The corresponding sides are proportional.
- Two pairs of corresponding sides are proportional, and the included angles are equal.

2.3 Right Triangle Properties

For the following properties, we refer to a right triangle *ABC* with its right angle at *C*, shown in Figure 2.1. The lengths of the sides opposite vertices *A*, *B*, and *C* are named *a*, *b*, and *c*, respectively. The altitude *CD* to the hypotenuse *AB* has length *h* and divides the hypotenuse into segments *AD* and *DB* with lengths *p* and *q*, respectively.

- The Pythagorean theorem: the sum of the squares of the legs equals the square of the hypotenuse: $a^2 + b^2 = c^2$.
- The square of the altitude *h* on the hypotenuse is equal to the product of the lengths of the segments *p* and *q* on the hypotenuse: $h^2 = p \cdot q$.
- The square of the length *a* of a leg is equal to the product of the length *c* of the hypotenuse and the adjoining altitude segment *q*: $a^2 = c \cdot q$, also $b^2 = c \cdot p$.
- The median *CM* to the hypotenuse of a right triangle is one-half the length *c* of the hypotenuse.

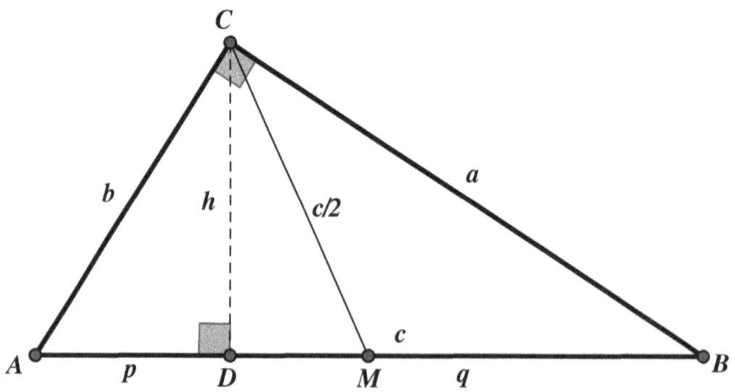

Figure 2.1

2.4 Angles Related to a Circle

These relationships are given in two different common notations, so as to accommodate the reader's background and also the applications be considered. Referring to Figure 2.2, we have the following circle-related relationships:

- An inscribed angle: $\angle A = \frac{1}{2}\widehat{BAC}$ (or $\angle CAB = \frac{1}{2} \cdot \angle COB$)
- An angle formed by 2 secants: $\angle F = \frac{1}{2}\left(\widehat{BC} - \widehat{DE}\right)$ (or $\angle CFB = \frac{1}{2} \cdot (\angle COB - \angle EOD)$)
- An angle formed by 2 tangents: $\angle J = \frac{1}{2}\left(\widehat{HAK} - \widehat{HCK}\right) = 180° - \widehat{HCK}$ (or $\angle HJK = \frac{1}{2} \cdot (180° - \angle KOH)$)
- An angle formed by a secant and a tangent: $\angle L = \frac{1}{2}\left(\widehat{CK} - \widehat{NK}\right)$ (or $\angle KLC = \frac{1}{2} \cdot (\angle KOC - \angle NOK)$)
- An angle formed by 2 chords intersecting inside the circle: $\angle AME = \frac{1}{2}\left(\widehat{AE} + \widehat{BC}\right)$ (or $\angle AME = \frac{1}{2} \cdot (\angle AOE + \angle COB)$)
- An angle formed by a chord and a tangent: $\angle JKC = \frac{1}{2}\angle\widehat{KC} = \angle KAC$ (or $\angle JKC = \frac{1}{2}\angle KOC = \angle KAC$)
- Opposite angles in a circle (i.e., opposite angles of an inscribed quadrilateral): $\angle BCK = 180° - \angle KAB$

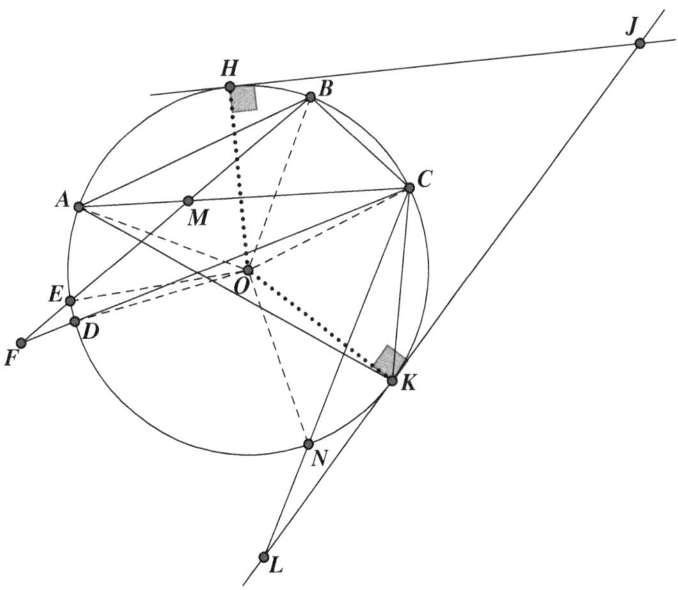

Figure 2.2

2.5 Tangents, Secants, and Chords of a Circle

The following refers to Figure 2.3:

- From the same external point, the tangent is a mean proportional between the entire secant and its external segment:

$$\frac{e+f}{h}=\frac{h}{f}.$$

- From the same external point, for two secants: the product of the length of a secant and its external segment is equal to the product of the other secant and its external segment: $(e+f)\cdot f=(g+k)\cdot k$.
- For two chords intersecting inside a circle: the product the segments of one chord is equal to the product or segments of the other chord: $a\cdot b=c\cdot d$.

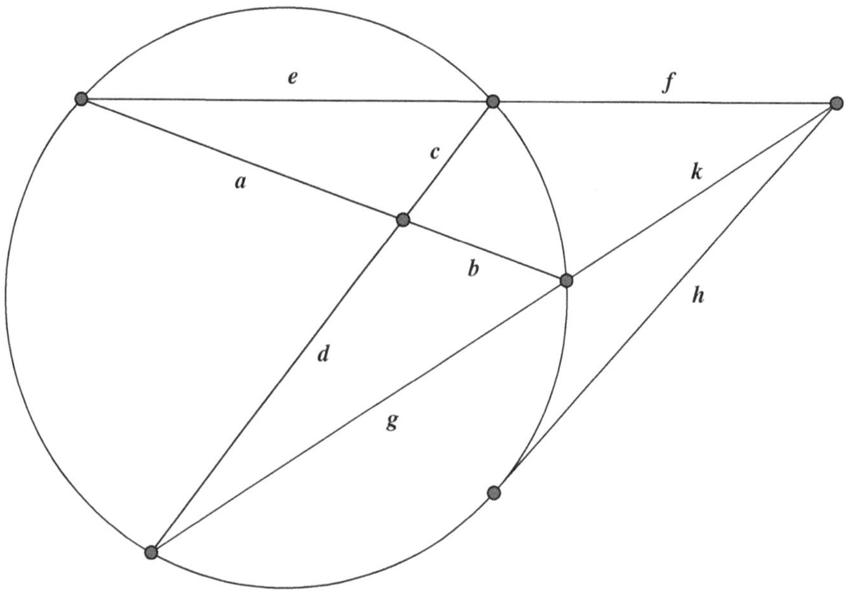

Figure 2.3

2.6 The Law of Sines and the Law of Cosines

Although the Law of Sines and the Law of Cosines do not occur very much in this book, we offer them here for the few times that they do occur. In triangle *ABC*, in Figure 2.4, we let *a*, *b*, and *c* denote the lengths of the sides opposite the vertices A, B, and C, respectively. Then

$$\frac{a}{\sin \angle A} = \frac{b}{\sin \angle B} = \frac{c}{\sin \angle C}.$$

This is known as the *Law of Sines.*

Also, $c^2 = a^2 + b^2 - 2ab\cos \angle ACB$, and analogously

$$a^2 = b^2 + c^2 - 2bc\cos \angle BAC$$

and

$$b^2 = c^2 + a^2 - 2ca\cos \angle CBA.$$

This is known as the *Law of Cosines.*

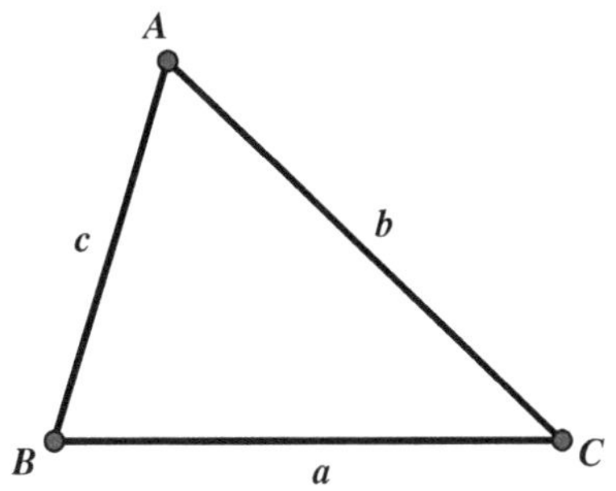

Figure 2.4

2.7 Trigonometry: Angle Sum and Difference Identities

It is often useful to express the sine or cosine of the sum or difference of two angles in terms of the individual sines and cosines. For this purpose, we can use the following identities:

$$\sin(\alpha + \beta) = \sin\alpha\cos\beta + \cos\alpha\sin\beta$$

$$\sin(\alpha - \beta) = \sin\alpha\cos\beta - \cos\alpha\sin\beta$$

$$\cos(\alpha + \beta) = \cos\alpha\cos\beta - \sin\alpha\sin\beta$$

$$\cos(\alpha - \beta) = \cos\alpha\cos\beta + \sin\alpha\sin\beta$$

2.8 Interior Angle Bisector in a Triangle

This relationship involving the triangle's angle bisector is not very common in the curriculum of the high school geometry course, so we present it here because can be very useful and we offer a simple proof. The interior angle bisector of a triangle partitions the side to which it is drawn proportional to the two adjacent sides of the bisected angle. In Figure 2.5, line AD is a bisector of angle $\angle BAC$, and therefore,

$$\frac{DC}{DB} = \frac{AC}{AB}.$$

A proof is offered for this not-too-well-known theorem.

Proof: In Figure 2.5, line AD is the bisector of $\angle BAC$ of ΔABC. The line parallel to AD through point B intersects the extension of side CA at point E. To make matters simpler, we have marked four angles in the figure, all of which are equal. We have $\angle ABE = \angle BAD$, as these are alternate-interior angles of the parallel lines AD and BE. Furthermore, $\angle BEA = \angle DAC$ as they are corresponding angles of these parallel lines,

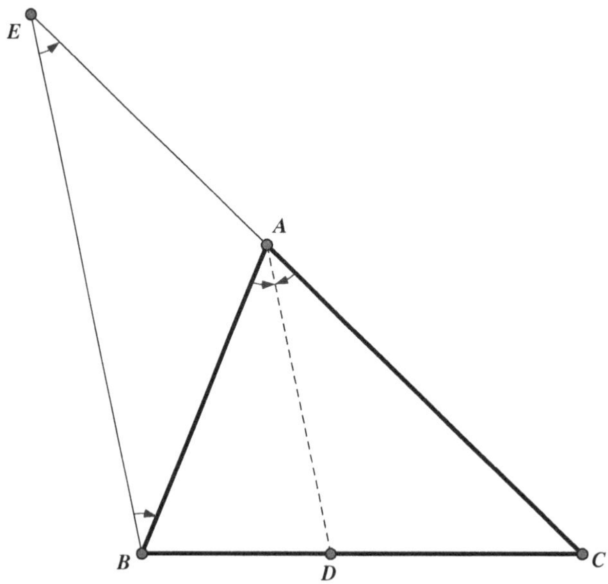

Figure 2.5

and since AD is the bisector of $\angle BAC$, we also have $\angle BAD = \angle DAC$. Therefore, by substitution we get $\angle BEA = \angle ABE$. This yields isosceles triangle ABE, where $AE = AB$. Since triangles EBC and ADC are similar, we have

$$\frac{DC}{DB} = \frac{AC}{AE}.$$

Now we substitute AB for AE to get the desired result, namely,

$$\frac{DC}{DB} = \frac{AC}{AB}.$$

2.9 Exterior Angle Bisector in a Triangle

The exterior angle bisector of a triangle partitions the line containing the opposite side to which it is drawn into segments proportional to the two adjacent sides of the bisected angle. In Figure 2.6, line AD is a bisector of angle $\angle EAB$, and therefore,

$$\frac{DC}{DB} = \frac{AC}{AB}.$$

Once again, a proof is provided to justify this powerful theorem.

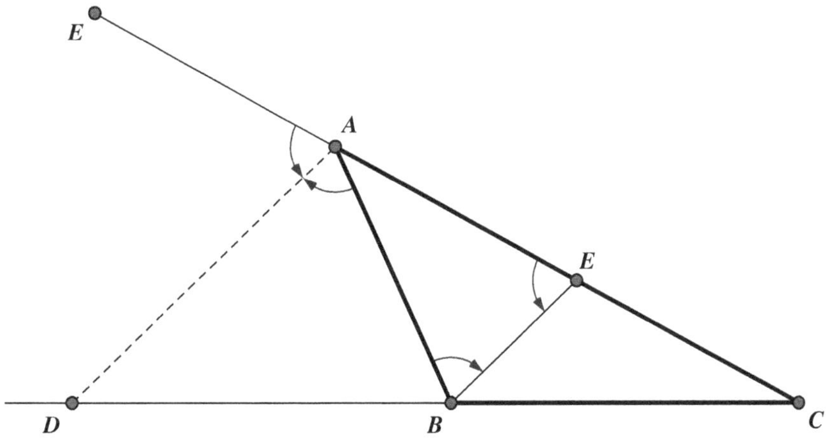

Figure 2.6

Proof: In Figure 2.6, line AD is the bisector of exterior angle $\angle EAB$ of ΔABC. The line parallel to AD through point B intersects side CA at point E. As in the previous proof, we have marked four angles in the figure, all of which are equal.

We have $\angle ABE = \angle DAB$, as these are alternate-interior angles of the parallel lines AD and BE. Furthermore, $\angle EAD = \angle AEB$ are corresponding angles of these parallel lines, and since AD is the bisector of $\angle EAB$, we also have $\angle EAD = \angle DAB$. Therefore, by substitution we get $\angle AEB = \angle EBA$. This yields isosceles triangle ABE, where $AE = AB$. Since triangles EBC and ADC are similar, we have

$$\frac{DC}{DB} = \frac{AC}{AE}.$$

We substitute AB for AE and once again get the desired result, namely,

$$\frac{DC}{DB} = \frac{AC}{AB}.$$

2.10 Some More Geometric Relationships Beyond the High School Geometry Course

The equilateral triangle presents a host of unusual and unexpected relationships. A number of them will be presented throughout the book. However, one very interesting relationship can be useful as we begin our tour through geometry.

Theorem: The sum of the distances from any point in the interior of an equilateral triangle to the sides of the triangle is constant, which is the length of the altitude of the triangle.

Two proofs of this interesting property are provided here. The first compares the length of each perpendicular segment to a portion of the altitude, and the second involves area comparisons.

Proof I: In Figure 2.7, we have equilateral ΔABC, with $PR \perp AC$, $PQ \perp BC$, $PS \perp AB$, and $AD \perp BC$. Draw a line through P parallel to BC, meeting AD, AB, and AC at G, E, and F, respectively.

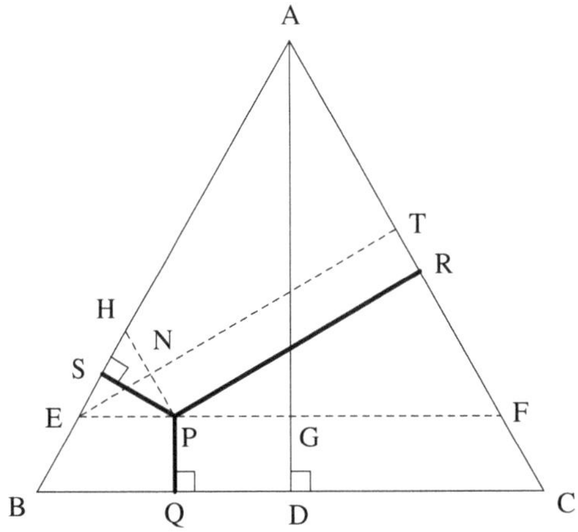

Figure 2.7

Since $PGDQ$ is a rectangle, $PQ = GD$. Draw $ET \perp AC$. Since ΔAEF is equilateral, $AG = ET$, as all the altitudes of an equilateral triangle are equal. Draw $PH \parallel AC$, intersecting ET at N so that $NT = PR$. Since ΔEHP is equilateral, altitudes PS and EN are equal. Therefore, we have shown that $PS + PR = ET = AG$. Since $PQ = GD$, we have $PS + PR + PQ = AG + GD = AD$, which is a constant for the given triangle.

Proof II: In Figure 2.8, the equilateral ΔABC has $PR \perp AC$, $PQ \perp BC$, $PS \perp AB$, and $AD \perp BC$. Then we draw PA, PB, and PC.

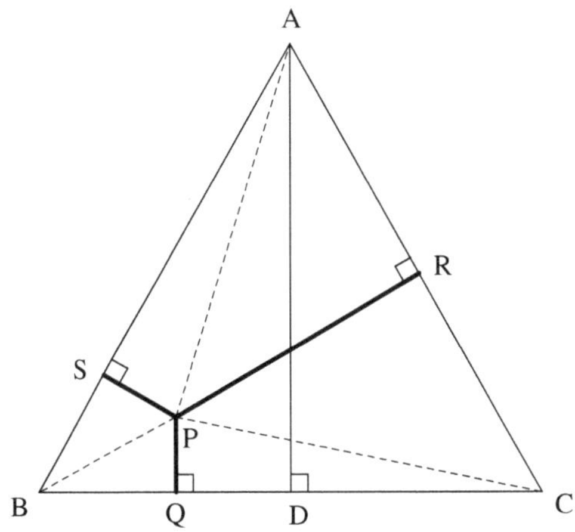

Figure 2.8

The area of ΔABC = area of ΔAPB + area of ΔBPC + area of $\Delta CPA = \frac{1}{2}(AB)(PS)+\frac{1}{2}(BC)(PQ)+\frac{1}{2}(AC)(PR)$.

Since $AB = BC = AC$, the area of $\Delta ABC = \frac{1}{2}(BC)[PS+PQ+PR]$.

However, the area of $\Delta ABC = \frac{1}{2}(BC)(AD)$. Therefore, $PS+PQ+PR=AD$, a constant for the given triangle.

There are times when it is useful to find the point in the quadrilateral with the sum of the distances to the vertices is a minimum. To prove that among the interior points of a quadrilateral, the point where the diagonals intersect has the smallest sum of distances to the vertices, we can choose any other interior point and compare its sum of distances to the vertices to that of the diagonal-intersection point.

Consider quadrilateral *ABCD* with diagonals *AC* and *BD* intersecting at *Q*, shown in Figure 2.9.

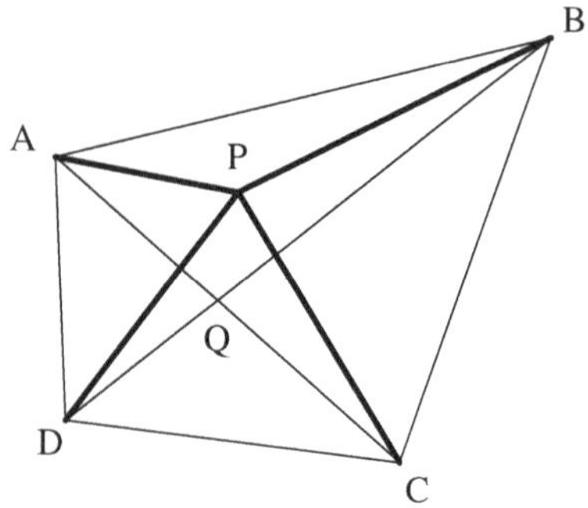

Figure 2.9

Select any point *P* (not at *Q*) in the interior of quadrilateral *ABCD*. Then *PA* + *PC* > *QA* + *QC*, since the sum of the lengths of two sides of a triangle is greater than the length of the third. Similarly, *PB* + *PD* > *QB* + *QD*. By addition, *PA* + *PB* + *PC* + *PD* > *QA* + *QB* + *QC* + *QD*, which shows that the sum of the distances from the point of intersection of the diagonals of a quadrilateral to the vertices is less than the sum of the distances from *any other* interior point of the quadrilateral to the vertices. This permits us to state the following theorem:

Theorem: The minimum-distance point of a quadrilateral is the point of intersection of the diagonals.

Next, we will consider where the minimum distance point of a triangle would be. We shall use ΔABC with no angle measuring greater than 120°, as shown in Figure 2.10.

Let M be the point in the interior of ΔABC, where $\angle AMB = \angle BMC = \angle AMC = 120°$. Draw lines through A, B, and C, which are perpendicular to AM, BM, and CM, respectively. These lines meet to form equilateral ΔPQR. To prove that ΔPQR is equilateral, notice that each angle has measure 60°. This can be shown by considering, for example, quadrilateral $AMBR$. Since $\angle RAM = \angle RBM = 90°$, and $\angle AMB = 120°$, it follows that $\angle ARB = 60°$. Let D be *any other* point in

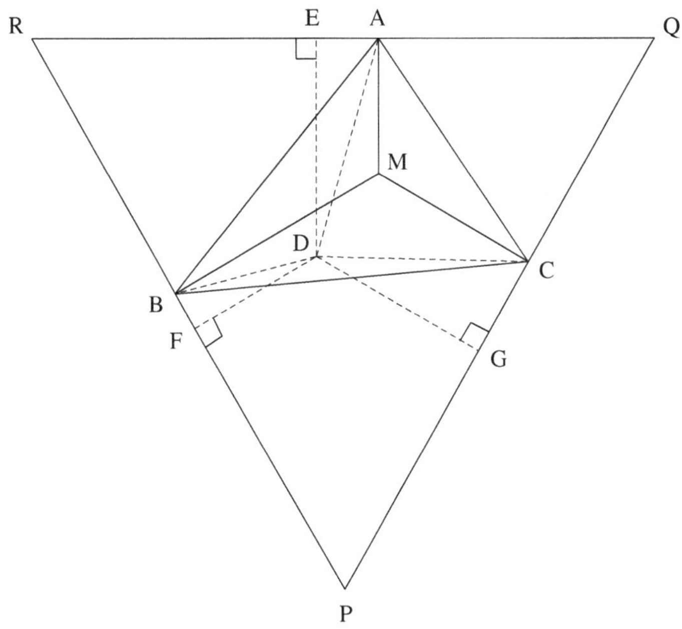

Figure 2.10

the interior of ΔABC. We must show that the sum of the distances from M to the vertices is less than the sum of the distances from D to the vertices.

Since we proved above that the sum of the distances from any point in the interior of an equilateral triangle to the sides of the triangle is a constant, we know that $MA + MB + MC = DE + DF + DG$, where DE, DF, and DG are the perpendiculars to REQ, RBP, and QGP, respectively.

But $DE + DF + DG < DA + DB + DC$. Recall that the shortest distance from an external point to a line is the length of the perpendicular segment from the point to the line. By substitution:

$$MA + MB + MC < DA + DB + DC.$$

We chose to restrict our discussion to triangles with angles of measure less than 120°. If you try to construct the point M in a triangle with one angle of measure of 150°, the reason for our restriction will become obvious. Thus, we have the following theorem.

Theorem: The minimum distance point of a triangle (with no angle of measure greater than 120°) is the equiangular point, which is the point at which the sides of the triangle subtend equal angles.

2.11 Heron's Formula for the Area of a Triangle

We know that when we are given the lengths of the three sides of a triangle, a unique triangle is determined. Therefore, in such a case we should be able to establish the area of the triangle. Heron of Alexandria (ca. 62 C.E.) developed a nifty formula to enable us to find the area of a triangle when the only information that we are given about a triangle is the lengths of its sides. This formula for the area of triangle ABC shown in Figure 2.11 is:

$$Area\,\Delta ABC = \sqrt{s(s-a)(s-b)(s-c)},$$

where $s = \dfrac{a+b+c}{2}$, which is the semi-perimeter of the triangle.

The development of this formula is provided further on in this chapter. Applying this formula is quite simple. For practice, we shall apply it to a triangle whose sides have lengths 13, 14, and 15 units, which then has a semi-perimeter of 21. Thus, with Heron's formula, we get the area of the triangle:

$$\sqrt{21\cdot(21-13)\cdot(21-14)\cdot(21-15)}=\sqrt{21\cdot8\cdot7\cdot6}=\mathbf{84.}$$

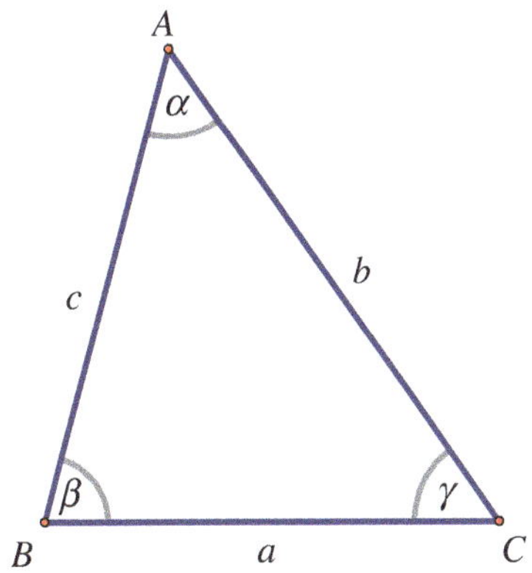

Figure 2.11

There are lots of other formulas for finding the area of a triangle, each requiring the measure of various parts of the triangle. We offer just some of these triangle-area formulas here, referring to Figure 2.12.

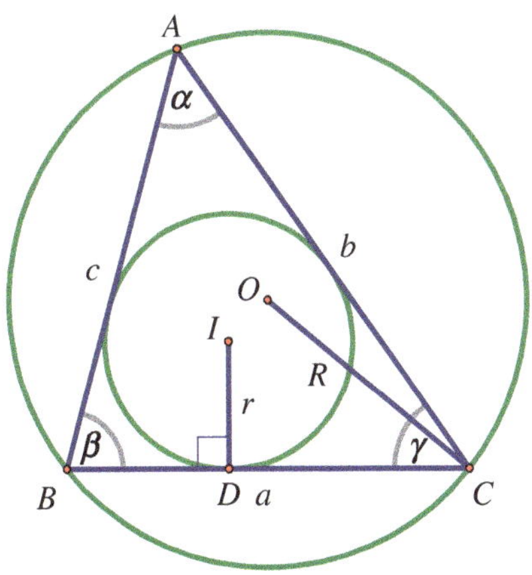

Figure 2.12

$Area\Delta ABC = \frac{abc}{4R}$ (R = radius of the circumscribed circle)

$Area\Delta ABC = r \cdot s$ (r = radius of the inscribed circle, and s = semi-perimeter)

$Area\Delta ABC = \frac{\tan\alpha}{4}\left(b^2 + c^2 - a^2\right)$ (when $\alpha \neq 90°$)

$Area\Delta ABC = \frac{a^2}{2} \cdot \frac{\sin\beta \cdot \sin\gamma}{\sin\alpha}$

$Area\Delta ABC = \frac{1}{4} \cdot \frac{a^2 + b^2 + c^2}{\cot\alpha + \cot\beta + \cot\gamma}$

$Area\Delta ABC = \frac{h_a h_b}{2\sin\gamma}$

$Area\Delta ABC = \frac{R \cdot h_a h_b}{C}$

$Area\Delta ABC = \frac{3}{4}\sqrt{m(m-m_a)(m-m_b)(m-m_c)}$, where m_a, m_b and m_c are the medians of ΔABC, and $m = \frac{m_a + m_b + m_c}{2}$.

2.12 Derivation of Heron's Formula for the Area of a Triangle

$$Area_{\Delta ABC} = \sqrt{s(s-a)(s-b)(s-c)},$$

where

$$s = \frac{a+b+c}{2}.$$

There is a nice geometric proof — actually attributed to Heron — provided in Thomas Heath's *A Manual of Greek Mathematics* (Dover, 1963). However, to conserve space, we will provide a much shorter trigonometric derivation building on the two high-school-level relationships: the Law of Cosines — $c^2 = a^2 + b^2 - 2ab \cos C$ — and the Pythagorean identity — $\sin^2\theta + \cos^2\theta = 1$. The area of a triangle formula is:

$$Area_{\Delta ABC} = \frac{1}{2}ab\sin C.$$

With $\cos^2 C = \dfrac{(a^2+b^2-c^2)^2}{4a^2b^2}$ and substituting for sin C in this equation for the area of a triangle, we get:

$$\begin{aligned} Area_{\Delta ABC} &= \frac{1}{2}ab\sin C = \frac{1}{2}ab\sqrt{1-\cos^2 C} \\ &= \frac{1}{2}ab\sqrt{1-\frac{(a^2+b^2-c^2)^2}{4a^2b^2}} \\ &= \frac{1}{2}ab\sqrt{\frac{4a^2b^2-(a^2+b^2-c^2)^2}{4a^2b^2}} \\ &= \frac{1}{4}\sqrt{4a^2b^2-(a^2+b^2-c^2)^2}. \end{aligned}$$

We now have to factor the term under the radical sign:

$$\begin{aligned} 4a^2b^2-(a^2+b^2-c^2)^2 &= -(a+b+c)\times(a+b-c)\times(a-b-c)\times(a-b+c) \\ &= -(a+b+c)\times(a+b-c)\times[-(-a+b+c)]\times(a-b+c) \\ &= (a+b+c)\times(a+b-c)\times(-a+b+c)\times(a-b+c). \end{aligned}$$

With $a+b+c=2s$, $a+b-c=2(s-c)$, $-a+b+c=2(s-a)$, and $a-b+c=2(s-b)$ we get

$$Area_{\Delta ABC} = \frac{1}{4}\sqrt{2s\cdot 2(s-c)\cdot 2(s-a)\cdot 2(s-b)} = \sqrt{s\cdot(s-a)\cdot(s-b)\cdot(s-c)}.$$

2.13 Brahmagupta's Theorem for the Area of a Quadrilateral

Now that you are familiar with Heron's famous formula for finding the area of any triangle given only the lengths of its three side, which is:

$$\text{Area of a triangle} = \sqrt{s(s-a)(s-b)(s-c)},$$

where a, b, and c are the lengths of the sides and $s = \dfrac{a+b+c}{2}$.

It is natural to try to extend this formula to quadrilaterals. One common way is to consider the triangle as a quadrilateral with a zero-length side. Such an extension is credited to Brahmagupta[1], an Indian mathematician who lived in the early part of the seventh century. Brahmagupta used the following formula to find the area of a *cyclic quadrilateral,* that is, a quadrilateral that may be inscribed in a circle, with sides of lengths a, b, c, and d, where s is the semi perimeter:

$$\text{Area} = \sqrt{(s-a)(s-b)(s-c)(s-d)}.$$

You will notice that Brahmagupta considered Heron's formula as treating the triangle as a quadrilateral with $d = 0$.

Proof of Brahmagupta's Formula: First consider the case where quadrilateral $ABCD$ is a rectangle with $a = c$ and $b = d$. Using Brahmagupta's formula, we have area of rectangle $ABCD$:

$$\begin{aligned}
&= \sqrt{(s-a)(s-b)(s-c)(s-d)} \\
&= \sqrt{(a+b-a)(a+b-b)(a+b-a)(a+b-b)} \\
&= \sqrt{a^2b^2} = ab
\end{aligned}$$

which is the area of the rectangle as found by the traditional methods.

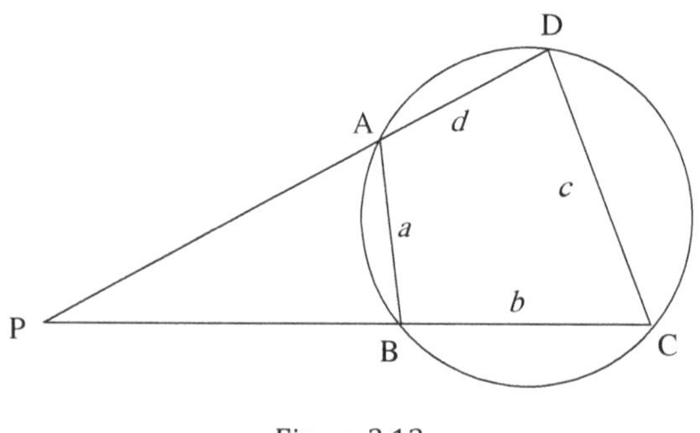

Figure 2.13

[1] In 628 Brahmagupta (born 598) wrote his *Brahma-sphuta-siddhānta* ("The Revised System of Brahma"), where the twelfth and thirteenth chapters are devoted to mathematics.

Consider any non-rectangular cyclic quadrilateral *ABCD*, as shown in Figure 2.13. Extend *DA* and *CB* to meet at *P*, forming ΔDCP. Let $PC = x$ and $PD = y$. By Heron's formula, area of

$$\Delta DCP = \frac{1}{4}\sqrt{(x+y+c)(y-x+c)(x+y-c)(x-y+c)}. \qquad \text{(I)}$$

Since $\angle CDA$ is supplementary to $\angle CBA$, and $\angle ABP$ is also supplementary to $\angle CBA$,

$$\angle CDA = \angle ABP. \text{ Then } \Delta BAP \sim \Delta DCP. \qquad \text{(II)}$$

From (II) we get

$$\begin{aligned}
&\frac{\text{area }\Delta BAP}{\text{area }\Delta DCP} = \frac{a^2}{c^2},\\
&\frac{\text{area }\Delta DCP}{\text{area }\Delta DCP} - \frac{\text{area }\Delta BAP}{\text{area }\Delta DCP} = \frac{c^2}{c^2} = \frac{a^2}{c^2},\\
&\frac{\text{area }\Delta DCP - \text{ area }\Delta BAP}{\text{area }\Delta DCP} = \frac{\text{area ABCD}}{\text{area }\Delta DCP} = \frac{c^2 - a^2}{c^2}.
\end{aligned} \qquad \text{(III)}$$

From (II) we also get

$$\frac{x}{c} = \frac{y-d}{a}, \qquad \text{(IV)}$$

and

$$\frac{y}{c} = \frac{x-b}{a}. \qquad \text{(V)}$$

By adding (IV) and (V),

$$\frac{x+y}{c} = \frac{x+y-b-d}{a}, \text{ and } x+y = \frac{c}{c-a}(b+d), \text{ so that}$$

$$x+y+c = \frac{c}{c-a}(b+c+d-a). \qquad \text{(VI)}$$

The following relationships are found by using similar methods.

$$y-x+c=\frac{c}{c+a}(a+c+d-b). \qquad \text{(VII)}$$

$$x+y-c=\frac{c}{c-a}(a+b+d-c). \qquad \text{(VIII)}$$

$$x-y+c=\frac{c}{c+a}(a+b+c-d). \qquad \text{(IX)}$$

Substitute (VI), (VII), (VIII), and (IX) into (I) to get the area of

$$\Delta DCP=\frac{c^2}{4\left(c^2-a^2\right)}\sqrt{(b+c+d-a)(a+c+d-b)(a+b+d-c)(a+b+c-d)}.$$

Since (III) may be read as the area of

$$\Delta DCP=\frac{c^2}{c^2-a^2}(\text{area } ABCD),$$

the area of cyclic quadrilateral

$$ABCD=\sqrt{(s-a)(s-b)(s-c)(s-d)}.$$

An interesting extension of Brahmagupta's formula to the general quadrilateral is given without proof. The area of any (convex)

$$\text{quadrilateral}=\sqrt{(s-a)(s-b)(s-c)(s-d)-abcd\cdot\cos^2\left(\frac{\alpha+\gamma}{2}\right)},$$

where a, b, c, and d are the lengths of the sides,

$$s=\frac{a+b+c+d}{2},$$

and α and γ are the measures of a pair of opposite angles of the quadrilateral. This formula shows that of all quadrilaterals that can be formed from four given side lengths, the one with the maximum area is the cyclic quadrilateral. The maximum area is achieved when

$$abcd\cdot\cos^2\left(\frac{\alpha+\gamma}{2}\right)=0,$$

which occurs when $\alpha + \gamma = 180°$, a fact true only in cyclic quadrilaterals.

Brahmagupta also found that for a cyclic quadrilateral of consecutive sides lengths *a, b, c,* and *d,* where *m* and *n* are the lengths of the diagonals the following relationship holds true:

$$m^2 = \frac{(ab+cd)(ac+bd)}{ad+bc}$$
$$n^2 = \frac{(ac+bd)(ad+bc)}{ab+cd}.$$

2.14 Ptolemy's Theorem for Cyclic Quadrilaterals

Perhaps the most famous theorem involving cyclic quadrilaterals is attributed to Claudius Ptolemaeus of Alexandria (c.100–170 CE), commonly known as Ptolemy. In his major astronomical work, the *Almagest*[2] (ca. 150 CE), he states this theorem on cyclic quadrilaterals.

Ptolemy's Theorem: The product of the lengths of the diagonals of a cyclic quadrilateral equals the sum of the products of the lengths of the pairs of opposite sides.

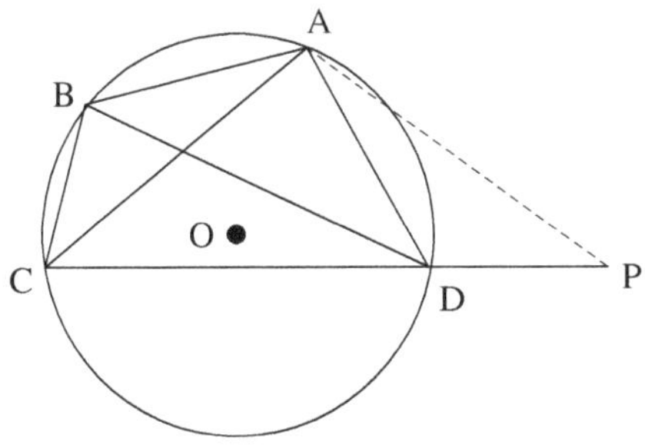

Figure 2.14

[2] The Greek title, *Syntaxis Mathematica,* means mathematical (or astronomical) compilation. The Arabic title, *Almagest,* is a renaming meaning "great collection (or compilation)." The book is a manual of all of the mathematical astronomy that the ancients knew at that time. Book I of the thirteen books that comprise this monumental work contains the theorem (6.11) that now bears Ptolemy's name.

Two methods for proving Ptolemy's theorem are provided. The second method incorporates the proof of the converse as well.

Ptolemy's Theorem Proof 1: In Figure 2.14, quadrilateral $ABCD$ is inscribed in circle O. A line is drawn through A to meet CD at P so that

$$\angle BAC = \angle DAP. \tag{I}$$

Since quadrilateral $ABCD$ is cyclic, $\angle ABC$ is supplementary to $\angle ADC$. However, $\angle ADP$ is also supplementary to $\angle ADC$.

Therefore,

$$\angle ABC = \angle ADP. \tag{II}$$

Thus,

$$\Delta BAC \sim \Delta DAP \text{ (AA)}, \tag{III}$$

and

$$\frac{AB}{AD} = \frac{BC}{DP}, \text{ or } DP = \frac{(AD)(BC)}{AB}. \tag{IV}$$

From (I), $\angle BAD = \angle CAP$, and from (III),

$$\frac{AB}{AD} = \frac{AC}{AP}.$$

Therefore, $\Delta ABD \sim \Delta ACP$ (SAS), and

$$\frac{BD}{CP} = \frac{AB}{AC}, \text{ or } CP = \frac{(AC)(BD)}{AB}, \tag{V}$$

so that

$$CP = CD + DP. \tag{VI}$$

Substituting (IV) and (V) into (VI),

$$\frac{(AC)(BD)}{AB} = CD + \frac{(AD)(BC)}{AB}.$$

Thus, $(AC)(BD) = (AB)(CD) + (AD)(BC)$.

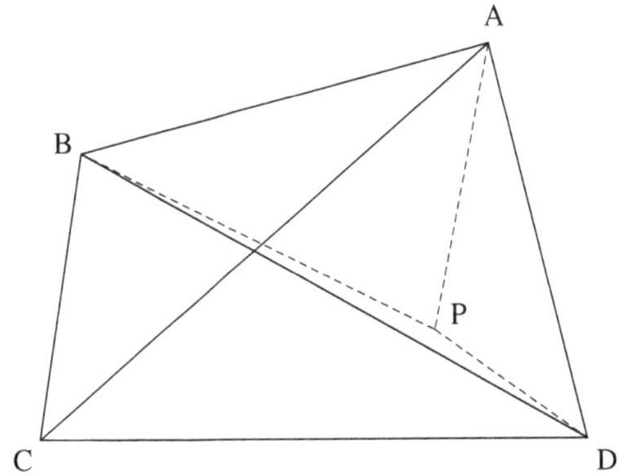

Figure 2.15

Ptolemy's Theorem Proof 2: In quadrilateral *ABCD,* as shown in Figure 2.15, draw ΔDAP on side *AD* similar to ΔCAB. Thus,

$$\frac{AB}{AP} = \frac{AC}{AD} = \frac{BC}{PD} \tag{I}$$

and

$$(AC)(PD) = (AD)(BC). \tag{II}$$

Since $\angle BAC = \angle PAD$, then $\angle BAP = \angle CAD$. Therefore, from (I), $\Delta BAP \sim \Delta CAD$ (SAS), and

$$\frac{AB}{AC} = \frac{BP}{CD}, \text{ or } (AC)(BP) = (AB)(CD). \tag{III}$$

Adding (II) and (III), we have

$$(AC)(BP + PD) = (AD)(BC) + (AB)(CD). \tag{IV}$$

Now $BP + PD > BD$ (triangle inequality), unless P is on DB. However, P will be on DB if, and only if, $\angle ADP = \angle ADB$. But we already know that $\angle ADP = \angle ACB$ (similar triangles). And if *ABCD* were cyclic, then $\angle ADB$ would equal $\angle ACB$, and $\angle ADB$ would equal $\angle ADP$. Therefore, we can state that, if and only if, *ABCD* is cyclic, P lies on BD.

This tells us that

$$BP + PD = BD. \tag{V}$$

Substituting (V) into (IV), $(AC)(BD) = (AD)(BC) + (AB)(CD)$.

Notice we have proved Ptolemy's Theorem and its converse, which we now state as our next theorem.

Converse of Ptolemy's Theorem: If the product of the lengths of the diagonals of a quadrilateral equals the sum of the products of the lengths of the pairs of opposite sides, then the quadrilateral is cyclic.

Assume quadrilateral $ABCD$ is not cyclic (see Figure 2.14). If there is a line CDP, then $\angle ADP \neq \angle ABC$. If C, D, and P are not collinear then it is possible to have $\angle ADP = \angle ABC$.

However, then $CP < CD + DP$ and from steps (IV) and (V), Proof I, above,

$$(AC)(BD) < (AB)(CD) + (AD)(BC).$$

But this contradicts the given information that $(AC)(BD) = (AB)(CD) + (AD)(BC)$. Therefore, quadrilateral $ABCD$ is cyclic.

2.15 Ceva's Theorem on Concurrency in a Triangle

We now consider one of the strongest theorems determining concurrency in a triangle. The theorem is often attributed to Giovanni Ceva (164–1734), who published it in his 1678 work *De lineis rectis se inviceem secantibus statica constructio*. But it is believed to have been proven much earlier by Yusuf Al-Mu'taman ibn Hűd, an eleventh-century king of Zaragoza.

Ceva's Theorem: The three lines containing the vertices A, B, and C of ΔABC and intersecting the opposite sides at points L, M, and N, respectively, are concurrent if, and only if,

$$\frac{AN}{NB} \cdot \frac{BL}{LC} \cdot \frac{CM}{MA} = 1.$$

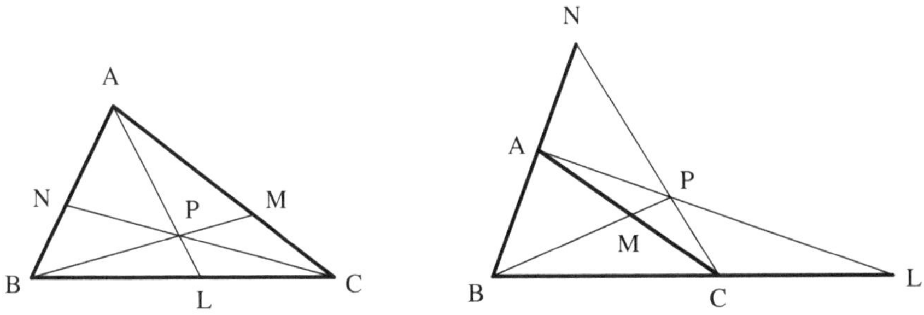

Figure 2.16

To prove Ceva's theorem, we must note that there are two possible situations in which the three lines drawn from the vertices may intersect the sides and still be concurrent, as shown in Figure 2.16. It is perhaps easier to follow the proof with the left-side diagram and verify the validity of the statements in the right-side diagram. In any case, the statements made in the proof hold for *both* diagrams.

Ceva's theorem is an equivalence (or biconditional) and, therefore, requires two proofs (one the converse of the other). We shall first prove that if the three lines containing the vertices of ΔABC and intersecting the opposite sides in points L, M, and N, respectively, are concurrent, then

$$\frac{AN}{NB}\cdot\frac{BL}{LC}\cdot\frac{CM}{MA}=1.$$

Three proofs are offered. The first (although not the simplest) requires no auxiliary lines.

Ceva's Theorem Proof 1: In Figure 2.16, AL, BM, and CN meet in point P. Since ΔABL and ΔACL share the same altitude (i.e., from point A),

$$\frac{area\ \Delta ABL}{area\ \Delta ACL}=\frac{BL}{LC}. \qquad \text{(I)}$$

Similarly,

$$\frac{area\ \Delta PBL}{area\ \Delta PCL}=\frac{BL}{LC}. \qquad \text{(II)}$$

From (I) and (II):

$$\frac{area\ \Delta ABL}{area\ \Delta ACL} = \frac{area\ \Delta PBL}{area\ \Delta PCL}.$$

A basic property of proportions $\left(\frac{w}{x} = \frac{y}{z} = \frac{w-y}{x-z}\right)$ provides that:

$$\frac{BL}{LC} = \frac{area\ \Delta ABL\text{-}area\ \Delta PBL}{area\ \Delta ACL\text{-}area\ \Delta PCL} = \frac{area\ \Delta ABP}{area\ \Delta ACP}. \qquad \text{(III)}$$

We now repeat the same process as above using *BM* instead of *AL* (used above).

Here:

$$\frac{CM}{MA} = \frac{area\ \Delta BMC}{area\ \Delta BMA} = \frac{area\ \Delta PMC}{area\ \Delta PMA}.$$

It follows that:

$$\frac{CM}{MA} = \frac{area\ \Delta BMC\text{-}area\ \Delta PMC}{area\ \Delta BMA\text{-}area\ \Delta PMA} = \frac{area\ \Delta BCP}{area\ \Delta BAP}. \qquad \text{(IV)}$$

Once again, we repeat the process using *CN* where *AL* was used earlier.

$$\frac{AN}{NB} = \frac{area\ \Delta ACN}{area\ \Delta BCN} = \frac{area\ \Delta APN}{area\ \Delta BPN}.$$

This permits us to get:

$$\frac{AN}{NB} = \frac{area\ \Delta ACN\text{-}area\ \Delta APN}{area\ \Delta BCN\text{-}area\ \Delta BPN} = \frac{area\ \Delta ACP}{area\ \Delta BCP}. \qquad \text{(V)}$$

We now simply multiply (III), (IV), and (V) to get the desired result:

$$\frac{BL}{LC} \cdot \frac{CM}{MA} \cdot \frac{AN}{NB} = \frac{area\ \Delta ABP}{area\ \Delta ACP} \cdot \frac{area\ \Delta BCP}{area\ \Delta BAP} \cdot \frac{area\ \Delta ACP}{area\ \Delta BCP} = 1.$$

By introducing an auxiliary line, we can produce a simpler proof.

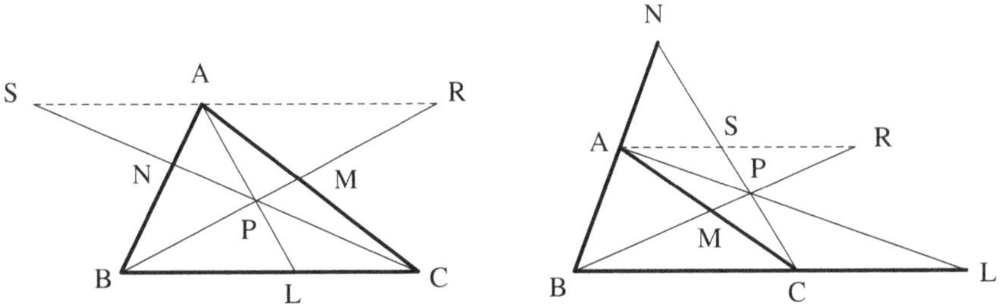

Figure 2.17

Ceva's Theorem Proof 2: Consider Figure 2.16 but include a line containing A and parallel to BC intersecting CP at S and BP at R, as shown in Figure 2.17. The parallel lines enable us to establish the following pairs of similar triangles:

$$\Delta AMR \sim \Delta CMB, \text{ therefore } \frac{AM}{MC} = \frac{AR}{CB} \tag{I}$$

$$\Delta BNC \sim \Delta ANS, \text{ therefore } \frac{BN}{NA} = \frac{CB}{SA} \tag{II}$$

$$\Delta CLP \sim \Delta SAP, \text{ therefore } \frac{CL}{SA} = \frac{LP}{AP} \tag{III}$$

$$\Delta BLP \sim \Delta RAP, \text{ therefore } \frac{BL}{RA} = \frac{LP}{AP}. \tag{IV}$$

From (III) and (IV) we get:

$$\frac{CL}{SA} = \frac{BL}{RA}.$$

This can be rewritten as

$$\frac{CL}{BL} = \frac{SA}{RA}. \tag{V}$$

Now by multiplying (I), (II), and (V) we obtain our desired result:

$$\frac{AM}{MC}\cdot\frac{BN}{NA}\cdot\frac{CL}{BL}=\frac{AR}{CB}\cdot\frac{CB}{SA}\cdot\frac{SA}{RA}=1.$$

We rearrange the terms and invert the fractions to get:

$$\frac{AN}{NB}\cdot\frac{BL}{LC}\cdot\frac{CM}{MA}=1.$$

(This reads the same as the conclusion of Proof 1).

By adding two auxiliary lines to the diagrams in Figure 2.16, we are able to produce another proof, again using the properties of similar triangles.

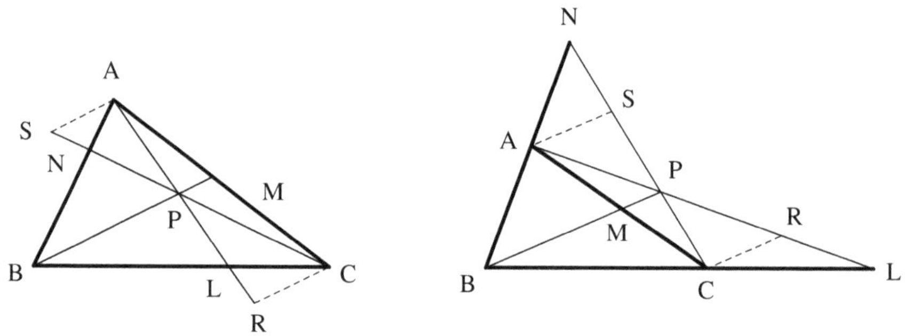

Figure 2.18

Ceva's Theorem Proof 3: We begin with the diagrams shown in Figure 2.9 but add two lines to each. We draw a line through A and a line through C, each parallel to BP and intersecting CP and AP at points S and R, respectively, as seen in Figure 2.18.

$$\Delta ASN \sim \Delta BPN, \text{ which yields } \frac{AN}{NB}=\frac{AS}{BP} \qquad \text{(I)}$$

$$\Delta BPL \sim \Delta CRL, \text{ which yields } \frac{BL}{LC}=\frac{BP}{CR} \qquad \text{(II)}$$

$$\Delta PAM \sim \Delta RAC, \text{ which yields } \frac{CA}{MA} = \frac{RC}{PM}.$$

This can be written as:

$$CA = \frac{(RC)(MA)}{PM}. \tag{III}$$

$$\Delta PCM \sim \Delta SCA, \text{ which yields } \frac{CM}{CA} = \frac{PM}{AS}.$$

This can be written as:

$$CA = \frac{(AS)(CM)}{PM}. \tag{IV}$$

From (III) and (IV):

$$\frac{(RC)(MA)}{PM} = \frac{(AS)(CM)}{PM},$$

which then can be written as:

$$\frac{CM}{MA} = \frac{RC}{AS} \tag{V}$$

$$\frac{AN}{NB} \cdot \frac{BL}{LC} \cdot \frac{CM}{MA} = \frac{AS}{BP} \cdot \frac{BP}{CR} \cdot \frac{RC}{AS} = 1.$$

To complete the proof of Ceva's theorem, we must now prove the converse of the implication proved above, that is, we shall now prove that, if the lines containing the vertices of ΔABC intersect the opposite sides in points *L, M,* and *N*, respectively, so that

$$\frac{AN}{NB} \cdot \frac{BL}{LC} \cdot \frac{CM}{MA} = 1,$$

then these lines, AL, BM, and CN are concurrent.

Suppose BM and AL intersect at P. Draw PC and call its intersection with AB point N'. Now that AL, BM, and CN' are concurrent, we can use the part of Ceva's theorem proved earlier to state the following:

$$\frac{AN'}{N'B}\cdot\frac{BL}{LC}\cdot\frac{CM}{MA}=1.$$

But our hypothesis stated that:

$$\frac{AN}{NB}\cdot\frac{BL}{LC}\cdot\frac{CM}{MA}=1.$$

Therefore,

$$\frac{AN'}{N'B}=\frac{AN}{NB},$$

so that N and N' must coincide, and thereby, prove concurrency.

Now that we have covered the basics of geometry and then some, we are ready to appreciate the geometry that is perhaps beyond the typical high school curriculum, but certainly within easy reach of a motivated reader.

Chapter 3

Euclid's Fifth Postulate: Various Models of Geometry

What is geometry? The answer may lie in the famous work the *Elements*, by the Greek mathematician Euclid of Alexandria (fl. 300 BCE), who developed the earliest axiomatic system in mathematics in the field of geometry. In his *Elements*, a collection of 13 books — one of the greatest successes in world literature — he logically deduces geometric theorems and constructions from a small set of postulates and axioms. These are the same constructions and theorems we still encounter today in secondary school geometry. But we also know that the constructions of this type of geometry may not be useful for a pilot or an astronomer, since we live on a sphere and not on a flat world and, therefore, in some instances other laws of geometry are valid.

Throughout this chapter, we will focus on Euclid's fifth postulate, and its influence on a type of geometry, and take a journey to discover fascinating examples of elliptic and hyperbolic geometries that go beyond Euclid's fifth postulate.

3.1 The Fifth Postulate

Euclid establishes five postulates at the beginning of his first book:

1. A straight line may be drawn between any two points.
2. A piece of straight line may be extended indefinitely.
3. A circle may be drawn with any given radius and an arbitrary center.
4. All right angles are equal.
5. If a straight line intersecting two straight lines makes the interior angles on the same side less than two right angles, then the two straight lines, if produced indefinitely, will intersect on the side on which the angles are less than two right angles.

The first four postulates seem easily understandable and reasonable. In order to comprehend the meaning of the fifth postulate, we take a look at Figure 3.1: The straight line (a) intersecting two straight lines (b and c) makes the interior angles on the same side less than two right angles (marked in Figure 3.1), the two straight lines (b and c), if produced indefinitely (dashed), meet (at point P) on that side of a on which the angles are less than two right angles.

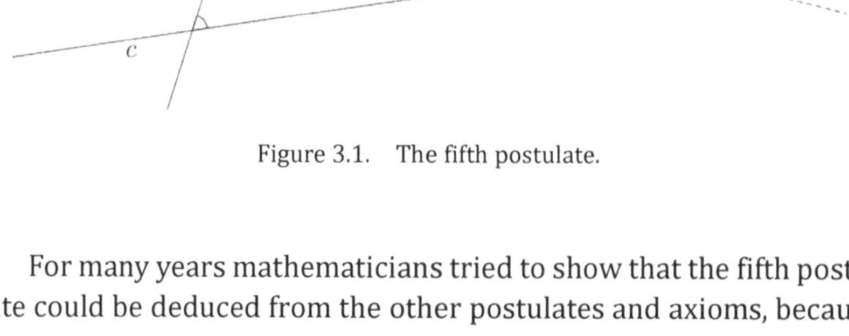

Figure 3.1. The fifth postulate.

For many years mathematicians tried to show that the fifth postulate could be deduced from the other postulates and axioms, because unlike the first four postulates, it is not self-evident. The only results they found were equivalent statements such as one of the best known by the Scottish mathematician John Playfair (1748–1819): "In a plane, given a line and a point not on it, at most one line parallel to the given line can be drawn through the point."

In Figure 3.2, we have Playfair's axiom pictured. Here, the given line is *a*, with the point *P* not on line *a*, and the parallel line is *b* (dashes).

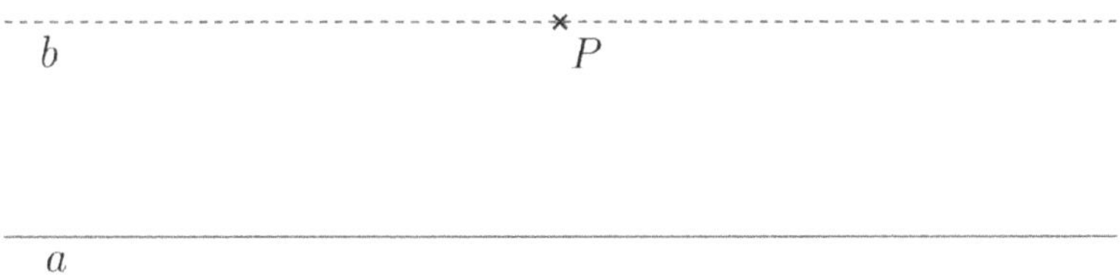

Figure 3.2. Playfair's axiom.

Another equivalent statement to Playfair's axiom is the triangle postulate: "The sum of the angles of a triangle is equal to two right angles." Given a triangle ΔABC in the Euclidean plane, Playfair's axiom guarantees that there exists exactly one parallel line *g* to the side *c*, which passes through the vertex *C* (Figure 3.3). If we now consider the theorem, that if two parallel lines *c* and *g* are cut by transversals *b* and *a*, then the alternate interior angles (dashed and dotted) are congruent, and we can confirm the angle sum of 180°.

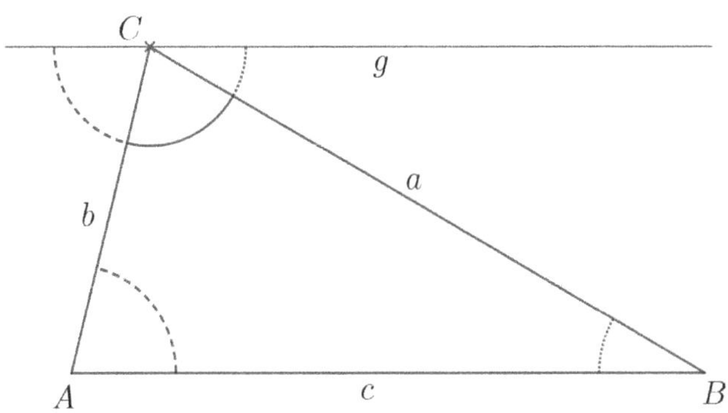

Figure 3.3. Sum of angles in a triangle in the Euclidean plane.

During the 2000 years while mathematicians were searching for a proof of Euclid's fifth postulate, they found many other logically equivalent formulations. The Russian mathematician Nikolai Ivanovich Lobachevsky (1792–1856), the Hungarian mathematician János Bolyai (1802–1860), and the German mathematician Carl Friedrich Gauss (1777–1855) tried to disprove Euclid's fifth postulate, but instead they found two "new types" of logically consistent geometries, that were later developed by Lobachevsky, the German mathematician Bernhard Riemann (1826–1866), and the French mathematician Henri Poincaré (1854–1912): these are hyperbolic and elliptic geometries. The Italian mathematician Eugenio Beltrami (1835–1900) finally demonstrated the independence of the parallel postulate from Euclid's other axioms. When we consider the sum of angles in a triangle, we get the three types shown in Figure 3.4.

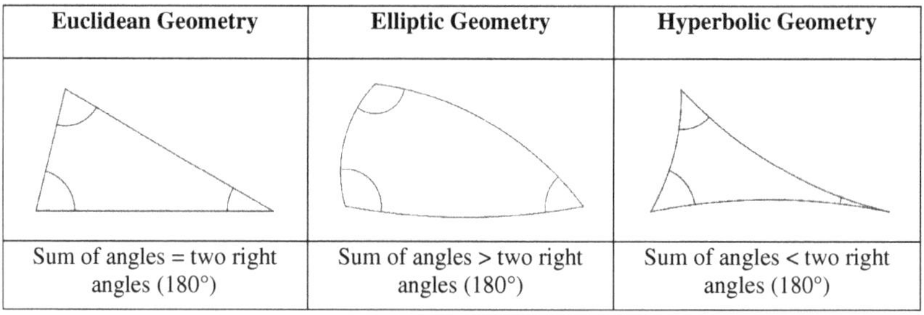

Figure 3.4. Sum of angles in a triangle.

What looks strange to us at first glance is that the sides of the two triangles on the right side in Figure 3.4 are not *straight* — but we will see in the following section that there exist also such triangles in our environment — the earth — and that it makes absolute sense to deal with them.

3.2 Geometry on a Sphere (Earth): A Model of an Elliptic Geometry

To accurately describe the location of a place on Earth, we use longitude and latitude. For example, Vienna, the capital of Austria, is located at 48.21° N latitude and 16.37° E longitude, as shown in Figure 3.5. We want to consider exactly how this location information is to be understood. If we now want to determine the position of Vienna on our globe, we have to measure from the zero meridian — that is, the meridian which is a great circle[1] at right angles to the Earth's equator, running from the North Pole N to the South Pole S, from which the geographical longitude is counted from east to west from the prime meridian (located in Greenwich, England) — 16.37°, and then from there 48.21° from the North Pole.

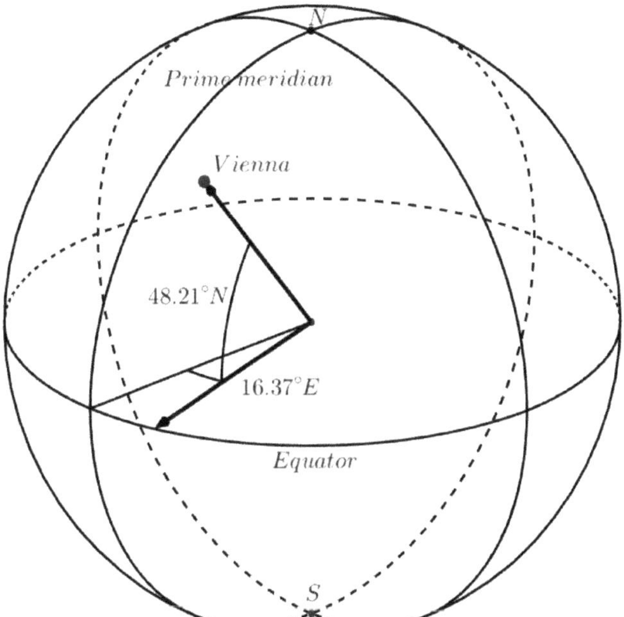

Figure 3.5. Latitude and longitude.

[1] The great circle of a sphere is the circle whose center is at the center of the sphere.

Although the Earth approximates a spherical shape, it is usually considered as a sphere. If one wants to solve geometrical problems that concern the Earth's surface, special geometry is used. For example, it is not possible to accurately determine the distance of two places that are not close to each other on the globe using plane Euclidean geometry. The shortest distance between two points on a sphere is an arc of a great circle. If we consider the triangle *ABN* in Figure 3.6, we can astonishingly determine that the sum of angles of ΔABN is equal to three right angles: 270°.

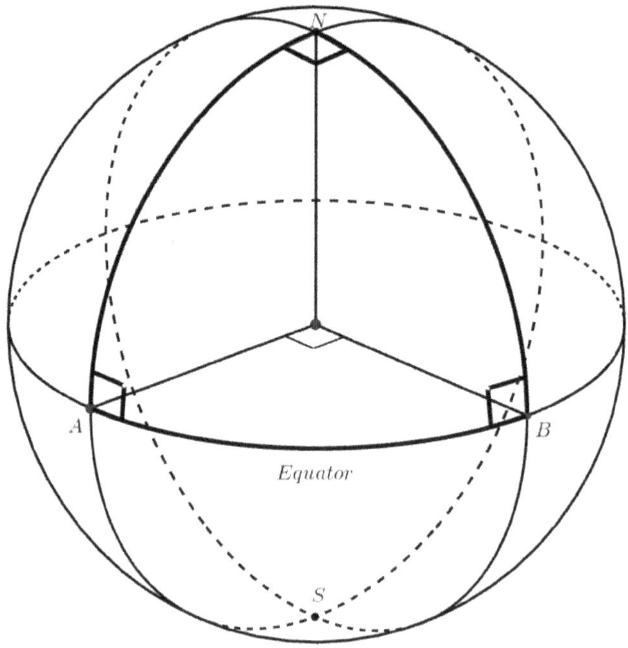

Figure 3.6. A triangle with the sum of angles of 270°.

Let's consider the maximum angle sum of a triangle on a sphere: If point *B* on the equator moves along the equator away from point *A*, the angle at *N* will eventually become 180°. Then the triangle "jumps" to the other side of the sphere if we agree that we do not allow reflex angles (the uniqueness of a triangle on a sphere will be lost). If point A moves away from point N along the longitude, point A will eventually arrive at point S. The angle in A will also eventually become 180° — as it becomes at point B, as shown in Figure 3.7.

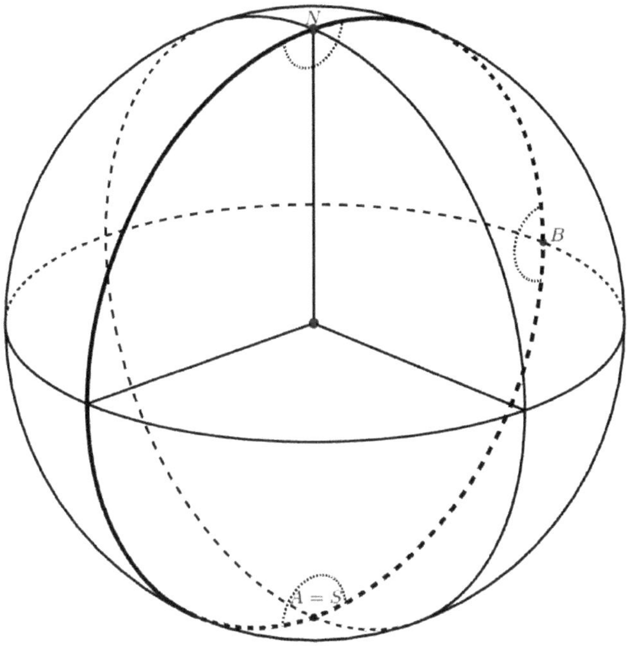

Figure 3.7. A "triangle" with the sum of angles of 540°.

The resulting geometric figure — according to plane Euclidean geometry — could no longer be called a triangle, since the three points all lie on a great circle, which in our Euclidean geometry would mean that they arc collinear. But if point B moves a little bit away from A, and also A does not move completely onto the South Pole S, we get a triangle which has an angular sum just less than 540°. For triangles small relative to the sphere, triangle *ABC* exhibits an angle sum only slightly larger than 180° (see Figure 3.8).

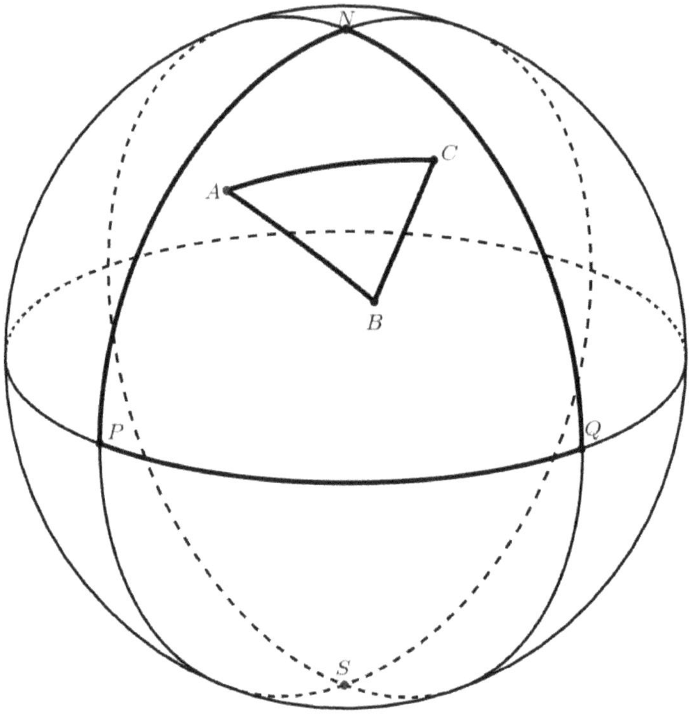

Figure 3.8. A smaller triangle on a sphere.

In summary, we can state that the sum of angles of a spherical triangle is between 180° and 540°, which is a big difference to Euclidean geometry, where the sum of the angles of a plane triangle is merely 180°. The study of triangles on spheres — as part of spherical trigonometry — yields many more differences with the Euclidean plane, which we will not consider here, as this would go beyond the scope of this book.

Instead, let's take a step back and try to think about how we can do geometry on a sphere in a meaningful way. Given two points *A* and

B on a sphere, we can always connect these two points with a great circle arc *c*, that is, one with the center of the circle at the center of the sphere, shown in Figure 3.9.

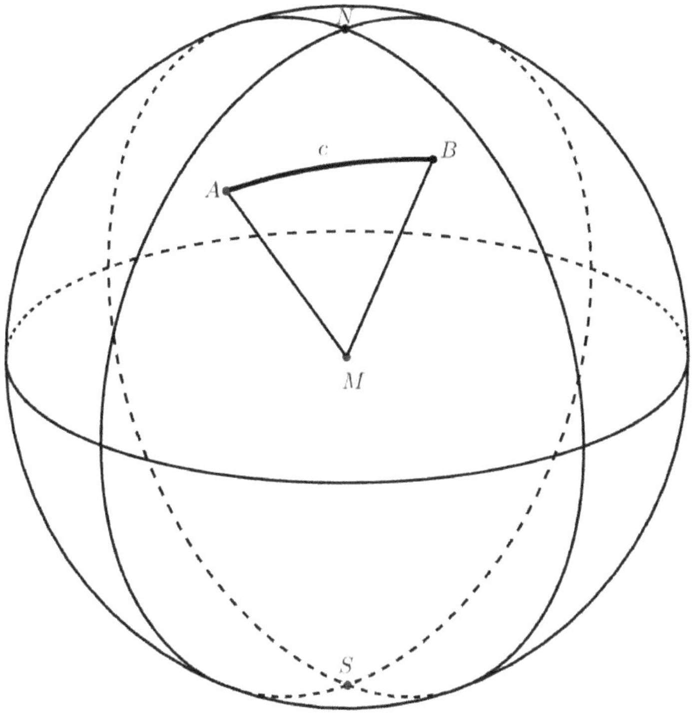

Figure 3.9. "Straight line" *c* from point *A* to point *B*.

The arc *c* fulfills the first postulate as a "straight line": One can draw a straight line from any point to any point. The second postulate is: One can extend a finite straight line continuously in a straight line. This is possible for our arc *c* as well — we can extend the arc on both sides as we like until we get the whole great circle with center *M* through *A* and *B*. The third postulate is not entirely valid in spherical geometry. On the surface of a sphere, all "circles" are actually great circles. A great circle always has the same radius as the radius of the sphere. This means that all great circles have the same radius. The fourth postulate, namely, that all right angles are equal to one another, is satisfied in spherical geometry in the sense of the definition of a right angle according to Euclid (Figure 3.10):

> *"If a straight line, placed on a straight line, forms equal subangles to each other, then each of the equal subangles is a right angle."*

This also holds for great circles on a sphere (Figure 3.11).

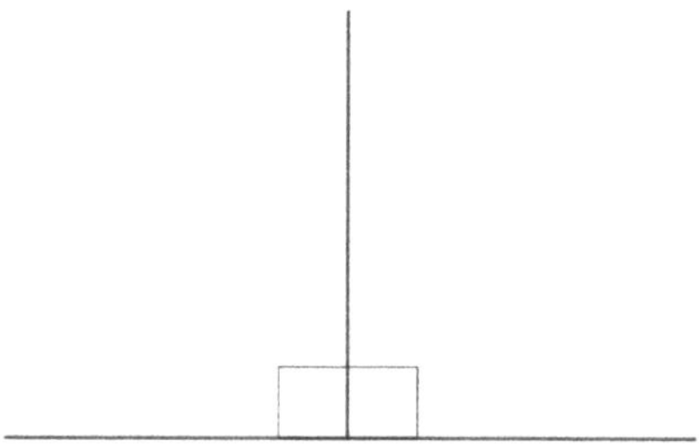

Figure 3.10. Right angles in Euclidean Plane.

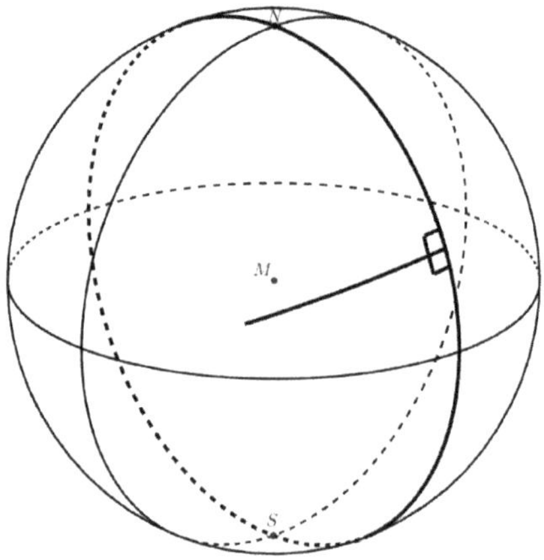

Figure 3.11. Right angles in spherical geometry.

Let us come back to the fifth postulate one more time. In John Playfair's version, he states:

> *"In a plane, given a line and a point not on it, at most one line parallel to the given line can be drawn through the point."*

The word "parallel" in Euclid's sense means that the straight lines have no common point. We can easily see that it's impossible to find such a parallel line on a sphere: To a given line c_1 and a given point P, every line c_2 through P intersects c_1 in two points because all lines are great circles on the sphere shown in Figure 3.12.

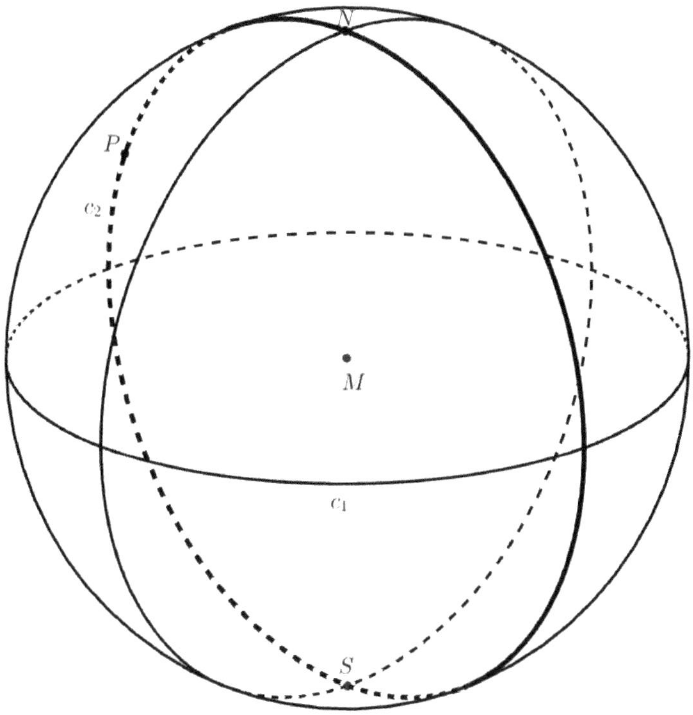

Figure 3.12. The non-existence of parallel lines on a sphere.

Since the Earth is almost a sphere, we should actually use the spherical geometry instead of the Euclidean geometry. However, we have seen that the differences only have an effect from a certain size of the figures in relation to the sphere; that is, when drawing geometric figures on a sheet of paper or when building a house, we can also work with Euclidean geometry with a clear conscience without making errors. But when we think, for example, of the shortest flight distance between two distant places on Earth (for example, between London and Los Angeles), we should bear in mind that this is always part of a great circle (Figure 3.13).

Figure 3.13. Approximately the shortest flight route between London and Los Angeles.

At first glance, the flight path in Figure 3.13 does not look like it is the shortest — however, if we try to find the shortest flight path on a globe rather than on a world map (see Figure 3.14), we see that it is in fact the shortest because, as we have already considered in this chapter, the shortest distance between two points on a sphere is an arc of a great circle. Since we cannot spread a sphere out into the plane without folding it, lengths and/or angles are distorted in a plane illustration, depending on the representation.

Another astonishing misinterpretation caused by plane representations of our world on maps is that the closest point of the USA to Africa must be in Florida. If we look closely at the situation on a globe, we realize that it is not in Florida at all, but in Maine. The distance between Maine and the Western Sahara is approximately 3,160 miles (see Figure 3.15).

Figure 3.14. Shortest flight route between London and Los Angeles (rubber band around a globe).

Figure 3.15. Shortest distance between the USA and Africa (rubber band around a globe).

3.3 The Poincaré Disk Model of a Hyperbolic Geometry

Now let's look at something completely different — some of the works of an artist who used geometric and mathematical knowledge to create his highly original and sometimes seemingly impossible images. Consider the works of the Dutch graphic artist Maurits Cornelis Escher (1898–1972).[2]

Among others, Escher created the series *Circle Limit* works, in which he explored hyperbolic tessellations. In *Circle Limit I* (1958), fish-like figures radiate from the center toward the boundary of the disk, becoming smaller and more densely packed. In *Circle Limit III* (1959), black and white fish-like figures are arranged symmetrically and shrink toward the boundary, giving the impression of infinite repetition. In *Circle Limit IV* (1960), also known as *Heaven and Hell*, angel and demon figures interlock seamlessly, filling the disk while diminishing in size toward the edge. These works illustrate how Escher visually employed the principles of hyperbolic geometry to create the effect of endless continuation within a finite circular frame.

In contrast to these artistic masterpieces, Figure 3.16 below shows only a simple schematic pattern of geodesics in the Poincaré disk. It is meant to illustrate the underlying geometric structure in an abstract and purely mathematical way, without the artistic richness of Escher's *Circle Limit* series.

Among other things, Escher dealt with tessellations, the gapless filling of the plane with figures. Escher explored the concept of hyperbolic geometry and used it to create tessellations that fill a circular disk with repeating patterns of interlocking shapes, often featuring angels, demons, or birds. Now let's look at the geometry behind these works of art.

The Poincaré disk model is a representation of hyperbolic geometry that maps the entire hyperbolic plane (which has infinite area) onto a disk in the Euclidean plane. This mapping preserves angles and shapes, but distances are distorted. The farther you move away from

[2] You can find the pictures and an extensive treatise on Escher's work in Doris Schattschneider, *Visions of Symmetry: Notebooks, Periodic Drawings, and Related Work of M. C. Escher*. W. H. Freeman, 1990.

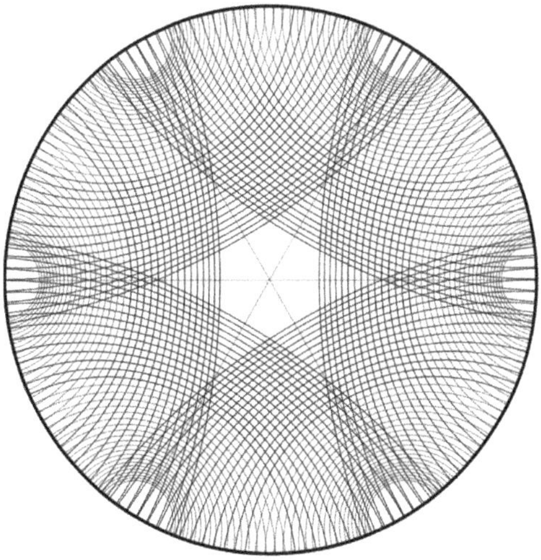

Figure 3.16. A simple schematic pattern of geodesics.

the center of the disk, the more distorted distances become. In the Poincaré disk model, lines are represented by hyperbolic geodesics, which are arcs of circles that intersect the boundary of the disk at right angles, shown in Figure 3.17. Hyperbolic angles are preserved by this model, and the disk's boundary is actually infinitely far away in hyperbolic terms.

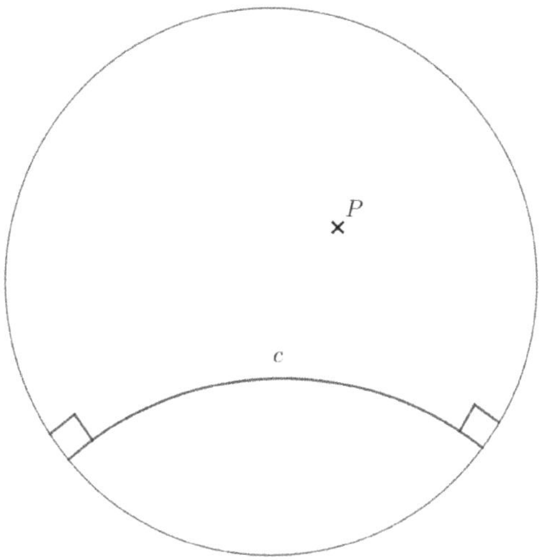

Figure 3.17. A line c and a point P in the Poincaré disk model.

We can construct the unique hyperbolic line through two points A and B (not on a diameter of the boundary circle) in the following way:

- We determine the image A' of point A by means of inversion in the boundary circle. Inversion is similar to reflection — both transformations involve mapping points across a boundary — but instead of a straight line, the boundary here is a circular shape (for more detailed information concerning this transformation see Chapter 6). We can obtain the image A′ of point A with the following construction steps (see Figure 3.18):
    - Draw ray r from the center M of the circle through A.
    - Draw line s through A perpendicular to r.
    - Let P be one of the two points where s intersects the circle.
    - Draw the segment PM.
    - Draw line t through P perpendicular to PM.
    - The image A' of point A is where ray r and line t intersect.

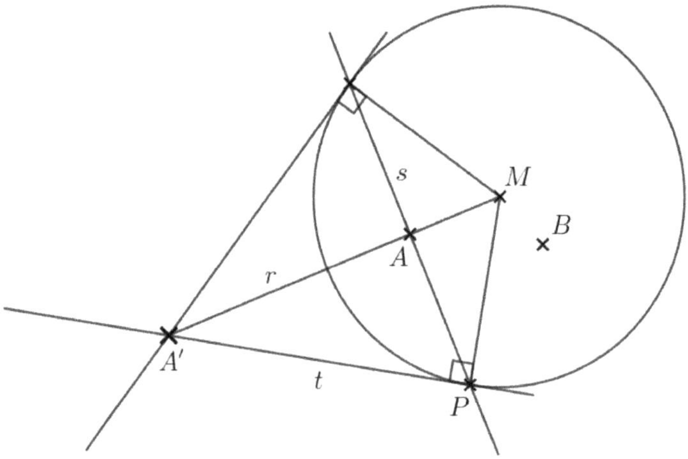

Figure 3.18. Construction of the inversion of point A.

- We determine the image B' of point B in the same way (for this and the following constructions see Figure 3.18).
- We determine the perpendicular bisectors m_A and m_B of AA' and BB'.

- The intersection of m_A and m_B provides a point C.
- We draw a circle through A, B with center C — the part of circle c that is inside the disk is the hyperbolic line we are looking for (see Figure 3.19).

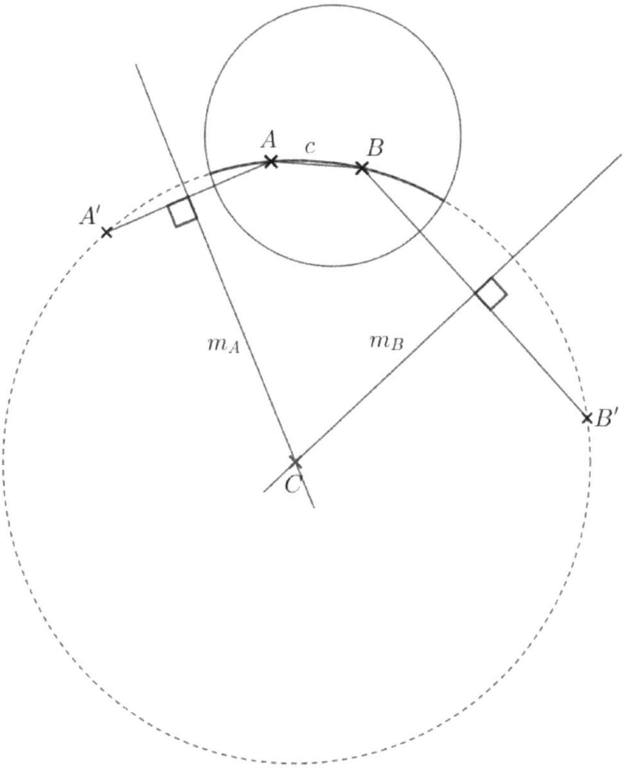

Figure 3.19. Construction of the hyperbolic connecting line through two points A and B.

If A and B are on a diameter of the boundary circle, that diameter is the hyperbolic line. The construction described above is far more complicated as compared to determining the connecting line in the Euclidean plane. Escher had to invest a lot of work in the creation of the series of pictures shown at the beginning of this chapter.

The term "parallel" in Euclid's sense means that the straight lines have no common point. We can easily find parallel lines in the

Poincaré disk: To a given line c and a given point P, there exists an infinite number of lines passing through point P without intersecting line c, for example, lines c_1, c_2, and c_3, which are shown in Figure 3.20.

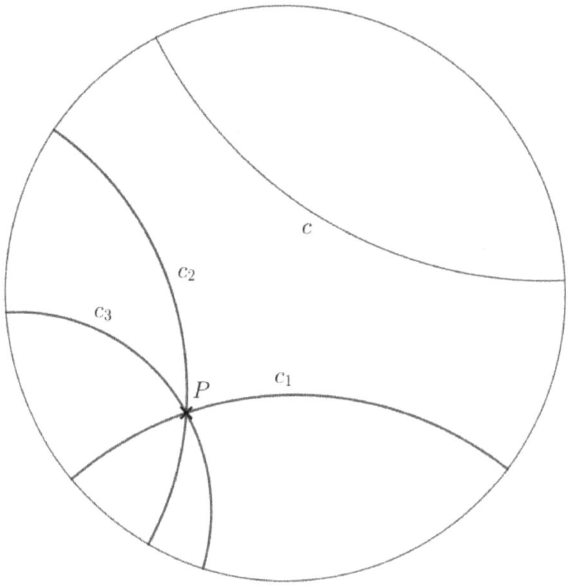

Figure 3.20. Parallel lines c_1, c_2 and c_3 through point P.

In addition to the diverse works of his artistic creation, consider seeing *M. C. Escher: Journey Into Infinity*, a documentary film from 2018.

Besides the art of M. C. Escher, hyperbolic geometry has numerous applications in various areas of mathematics, physics and computer science. In mathematics it is used in group theory, knot-theory, cryptography and discrete geometry. In physics it is deployed in the special and general theory of relativity, especially in the modeling of space-time curvatures. In computer graphics and visualization, hyperbolic geometry is utilized to generate non-Euclidean representations of geometric structures. This can be useful, for example, in the visualization of networks, social graphs or hierarchical structures. Let's imagine a network of social relationships, such as friendships in a community. In a conventional Euclidean representation, the nodes (people) and the edges (friendship relationships) would be arranged on a flat surface. This could result in certain parts of the network appearing crowded, while other parts appear spacious and disconnected. With hyperbolic geometry, however, you could represent this network on a curved, hyperbolic surface. On this curved surface, nodes with many connections will be close to each other, while isolated nodes could be far away. This representation would provide a better visualization of the actual social structure by emphasizing the actual proximity between people with many common friends, while highlighting the seclusion of individual people (Figure 3.21).

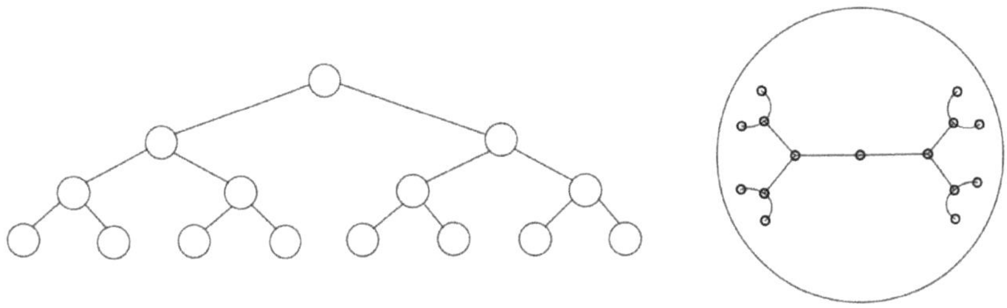

Figure 3.21. Euclidean vs. hyperbolic representation of a network.

In this chapter, we have come into contact with examples of three very different kinds of geometry and examples of their application. In the following chapters we will get to know a geometry that contains all of the three here-discussed kinds of geometries (Euclidean, elliptical, and hyperbolic geometries), which leads to valid statements for all of them: projective geometry.

Chapter 4

Innovative Geometric Techniques Using Projective Geometry

In this chapter, we will explore the fundamental concepts, methods, and findings of projective geometry. The development of projective geometry was a gradual process driven by various mathematicians. Some of the significant contributions came from perspectivists such as French mathematicians Girard Desargues (1591–1661) and Blaise Pascal (1623–1662). However, the more modern form of projective geometry was developed in the 19th century by German mathematicians such as Julius Plücker (1801–1861), Karl Georg Christian von Staudt (1798–1867), the French mathematician Jean-Victor Poncelet (1788–1867), and the Swiss mathematician Jakob Steiner (1796–1863), just to name a few.

The work of the 19th century mathematician Jakob Steiner is particularly important for the applications of fundamental concepts of projective geometry, which we will use throughout our study of the subject. His work contributed to the development of projective geometry into an independent field of mathematics. Some of Jakob Steiner's major contributions to projective geometry include the *principle of duality* of projectivity, which states that properties of points and lines in a point-line space are dual to each other, which means that in projective geometry, every theorem about points and lines remains true even if the terms "point" and "line" are consistently interchanged in that statement. We will discuss this important concept in detail in section 4.2 of this chapter and also look at some examples. He also introduced the Steiner ellipses, Steiner spheres, and the Steiner circle, which we will consider in Chapter 5.

For readers who wish to deepen their understanding or read further on the topic, the following books are recommended: Abraham Seidenberg, *Lectures in Projective Geometry*, Dover Publications, New York (2005); H. S. M. Coxeter, *Projective Geometry*, 2nd Ed., Springer, New York (1987). These sources provide rigorous and accessible treatments of the key ideas in projective geometry and are excellent references for further study.

4.1 Basic Axioms and Minimal Model

In the previous chapter, we explored the influence of Euclid's fifth postulate on the nature of geometric shapes, such as triangles, and saw how modifying this postulate leads to different types of geometry. We will now turn to a form of geometry that stands apart from these differences, a geometry that encompasses and unifies many others: *projective geometry*.

To develop this geometry, we begin in the most abstract and general way possible. We define a set of points $P = \{A, B, C, ...\}$, a set of lines $L = \{a, b, c, ...\}$, and a relation $I \subseteq P \times L$, called the *incidence relation*. This may sound technical at first, but the idea is simple: among all possible pairs $(X, y) \in P \times L$, only those that belong to I are the ones where the point X *lies on* the line y. In more intuitive terms, this relation captures the idea of a point being "on" a line or, conversely, a line "passing through" a point. These two perspectives are unified in the formal statement:

"The point is incident with the line."

In this framework, we do not define what points or lines "are" — we assign no physical or visual interpretation to them. Instead, their meaning arises entirely from the axioms that govern how they relate. As the German mathematician David Hilbert famously emphasized in the introduction to his foundational work *Foundations of Geometry* (1899):

> *"One must always be able to say, instead of 'points, lines, and planes,' 'tables, chairs, and beer mugs.'"*

This highlights the core idea behind the axiomatic method: mathematics is not about the objects themselves, but about the structure formed by their relationships.

Thus, projective geometry begins not with drawings or measurements, but with sets, relations, and a few simple axioms — offering a purely logical foundation that can accommodate a wide range of geometric systems.

To express incidence symbolically: If $(A, a) \in I$, we write $A \; I \; a$ (point A lies on line a); if $(A, a) \notin I$, we write $A \not{I} \, a$ (point A does not lie on line a). Using the two sets P (points), L (lines), and the relation I (incidence), we now define the projective plane by means of three basic axioms. The triple $\Pi = (P, L, I)$ is called a *projective plane* if the following three incidence axioms are satisfied:

(A1) Any two distinct points lie on exactly one line.

(A1) Any two distinct lines intersect at one point.

(A3) There exist at least four points, no three of which are collinear (there exists a quadrilateral).

Even if one might not arrive at such an example independently, it can be verified that the so-called *minimal model* indeed defines a projective plane of this kind (Figure 4.1).

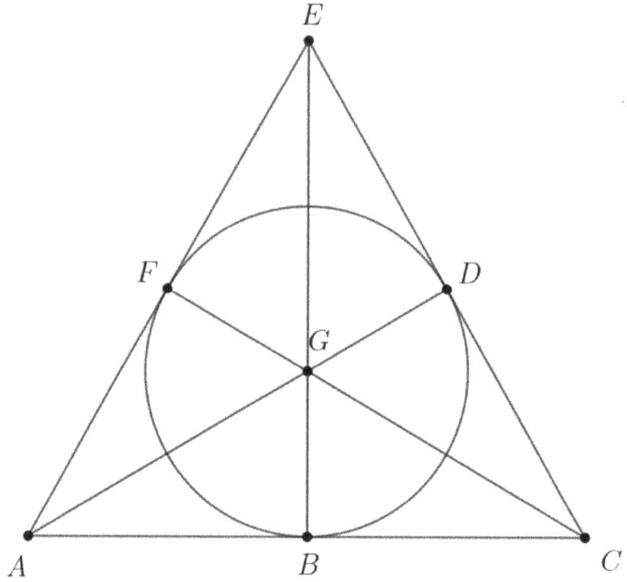

Figure 4.1. Minimal model.

When we look at the diagram in Figure 4.1, one line stands out — the one connecting the points *B, D,* and *F.* This line is, in fact, a circle. However, since our definition of a projective plane does not specify the shape a line must have, we can proceed without issue.

If we do not include the intersection points marked with circles in Figure 4.2 as part of our model, we obtain a structure consisting of exactly 7 points (*A, B, C, D, E, F, G*) and exactly 7 lines — namely, the lines connecting the point triples *ABC, CDE, AFE, AGD, CGF, BGE,* and *BDF.* Each line contains exactly three points, and through each point, exactly three lines pass.

If we go through all three axioms in order, we see that all of them are fulfilled in our model: every pair of two points lies on exactly one line, every pair of two lines intersects at exactly one point, and there exists a quadrilateral which could be the four points marked with circles in Figure 4.3.

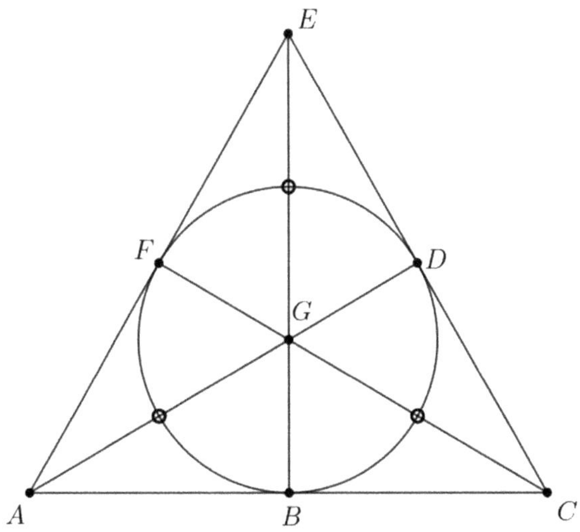

Figure 4.2. Excluded points.

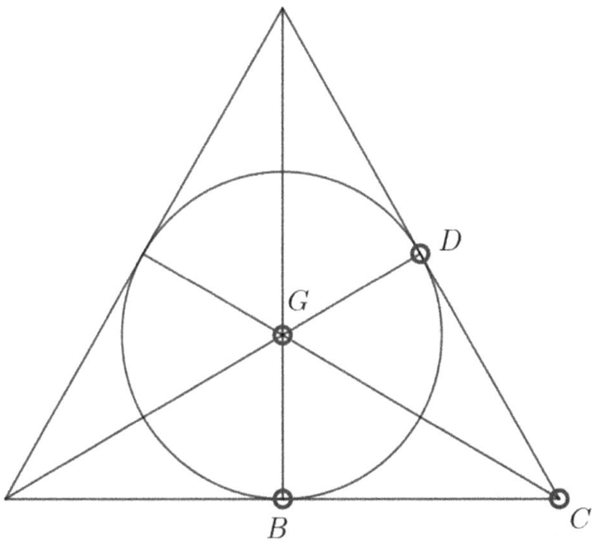

Figure 4.3. Quadrilateral.

Without going into the verification of axioms (A1)–(A3) at this point, it should be mentioned that the "next largest" finite projective plane consists of 13 points and 13 lines: 4 points lie on each line, and 4 lines pass through each point (see Figure 4.4).

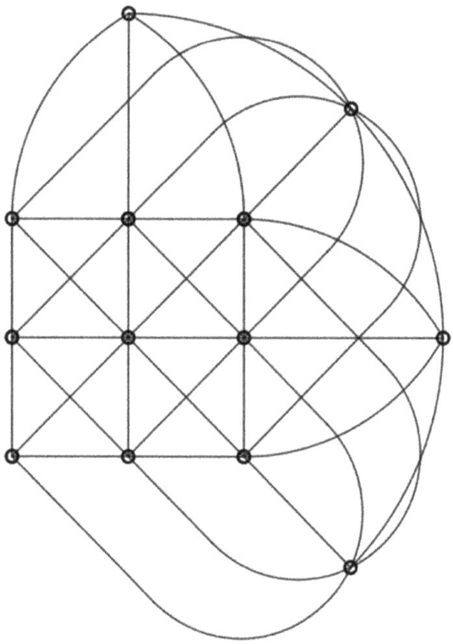

Figure 4.4. Model of a projective plane consisting of 13 points and 13 lines.

As a short look without diving too deeply, consider that when mathematicians try to classify finite projective planes, they often look at what is called the *order* of the plane. This is simply defined as the number of points on a line minus one. So, the smallest possible example — our minimal model — has order $3-1=2$, and the model shown in Figure 4.4 has order $4-1=3$.

But what numbers are actually possible as the order of a finite projective plane? That is still an open question. We do know that projective planes exist for all orders that are powers of a prime number, such as $4=2^2$, or $8=2^3$, or $5=5^1$, or $9=3^2$, and so on. But it is not clear whether those are the *only* possible orders. Interestingly, we do know that a projective plane of order 10 definitely does not exist — a major result that took decades to prove.

4.2 Duality

A very important principle in projective geometry is *duality*, which involves interchanging the words *point* and *line* while still maintaining a correct relationship. For example, *two points determine a unique line*, and *two intersecting lines determine a unique point*. This principle is not just a curiosity — it is reflected directly in the structure of the geometry itself. A closer look at axioms A1 and A2 shows that duality is built into the foundational rules.

(A1) Any two distinct points lie on exactly one line.

(A2) Any two distinct lines intersect at one point.

(A3) There exist at least four points, no three of which are collinear (there exists a quadrilateral).

Even the third axiom, A3, a so-called *richness axiom*, supports this symmetry: instead of postulating the existence of four non-collinear points, we could just as well assume the existence of four lines, no three of which are concurrent. Figure 4.5 illustrates this idea using the minimal model introduced earlier.

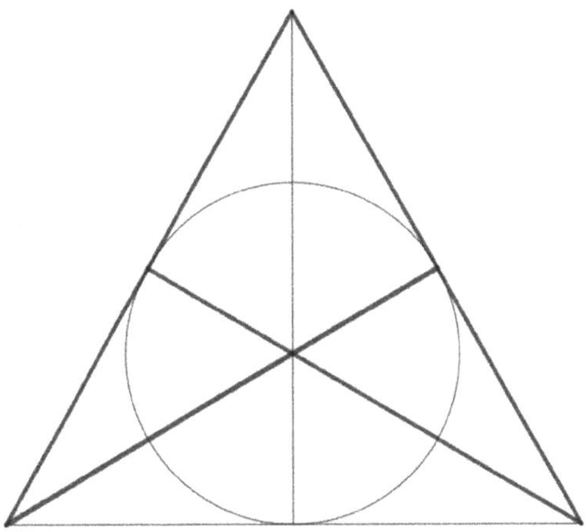

Figure 4.5. Four lines, no three of which are concurrent.

The duality principle has far-reaching implications in projective geometry — it allows us to take any proven theorem and interchange the words *point* and *line* to obtain another equally true theorem, without the need for a separate proof. This powerful symmetry means that for every theorem, there is a corresponding *dual* theorem that is automatically valid by virtue of duality. We will see other remarkable examples of this principle later on in this chapter.

4.3 The Projectively Completed (Euclidean) Plane

When we think about the incidence axioms (A1)–(A3), we notice that they are not always satisfied in the ordinary Euclidean plane. For example, parallel lines never meet — so they clearly do not intersect in exactly one point, as required by the axiom (A2) of projective geometry.

To resolve this, mathematicians imagine a *projectively completed version* of the Euclidean plane. This means we add so-called points at infinity, one for each family of parallel lines. All these points lie on a special line called the line at infinity. In this extended geometry, any two lines — even parallel ones — intersect in exactly one point. That point may lie in the usual plane or on the line at infinity. With this idea, the Euclidean plane becomes part of a broader system in which all lines meet and the projective incidence axioms are fully satisfied.

Figure 4.6 illustrates the situation for the parallel lines f, g, and h, which all meet at the point F on the line labeled ω, the line at infinity imagined to be infinitely far away.

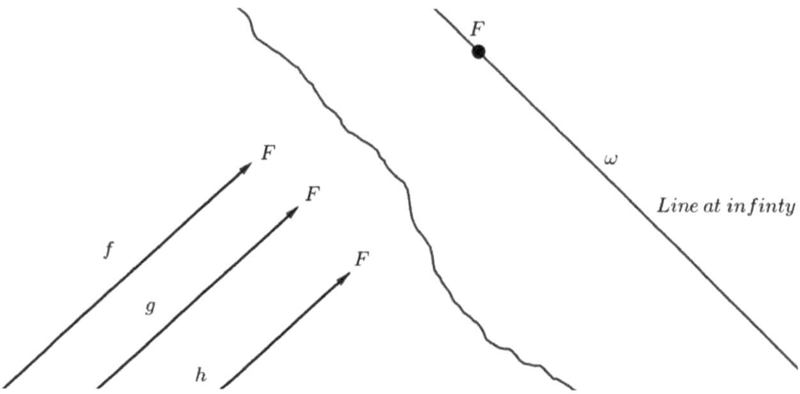

Figure 4.6. Parallel lines meeting at a point at infinity.

4.4 Perspectivities and Projectivities — Fundamental Mappings in Projective Geometry

In projective geometry, one of the most fascinating ideas is that geometric objects can be transformed while preserving their essential relationships — such as which points lie on which lines. The most fundamental types of such transformations are called *perspectivities* and *projectivities*.

A *perspectivity* is a simple transformation that mimics how we see the world. Imagine projecting points from one line a onto another b by looking through a fixed point — such as shining light from a lamp through a stencil onto a wall. The lamp L is the *center* of perspectivity, as shown in Figure 4.7. In mathematical notation, this is often written as:

$$a(X) \overset{L}{\underset{\wedge}{=}} b(Y).$$

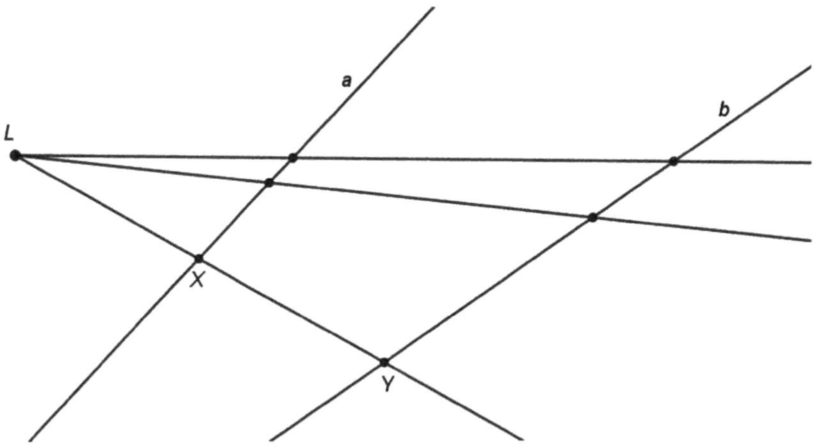

Figure 4.7. A perspectivity.

- It maps points from one line to another, always "through" a single point.
- It preserves alignments — if three points lie on a line before the transformation, they still lie on a line afterward.

In projective geometry, every concept has a dual — thanks to the powerful duality principle, which allows us to exchange the words "point" and "line" in any valid statement to get another valid one, as we stated earlier about the concept of duality.

Just as we can define a perspectivity by projecting points from one line through a center point to another line, we can also define a *dual perspectivity,* which maps lines through point A to lines through point B, with all corresponding pairs meeting on a common axis line l, as shown in Figure 4.8. This can be explained as follows:

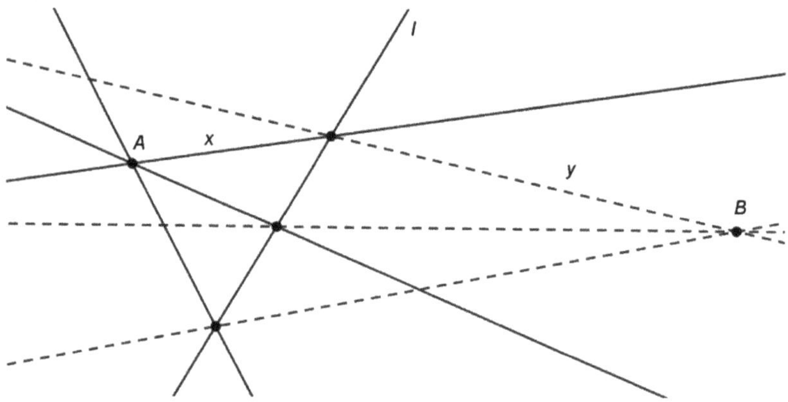

Figure 4.8. A dual perspectivity.

Consider two points, A and B, and a line l (the axis). Now imagine a series of lines going through A that also intersect l. For each of these points, draw the line that passes through the corresponding point of intersection on l and through B. This defines a mapping from the set of lines through A to the set of lines through B — and that is a dual perspectivity. In mathematical notation, this is often written as:

$$A(x) \overset{l}{\underset{\wedge}{=}} B(y).$$

Perspectivities form the building blocks of more complex transformations in projective geometry. While perspectivities are simple, elegant transformations based on a single center or axis, they are only the beginning. By combining several perspectivities, we unlock a much more powerful tool: the *projectivities.*

In projective geometry, a projectivity is a transformation that can be understood as a composition of perspectivities — the simplest kinds of projective mappings. A perspectivity maps points on one line to points on another through a fixed center point. By the duality principle, there is also a dual perspectivity, where lines through one point are mapped to lines through another point, such that all corresponding lines intersect on a fixed axis line. Combining several (primal or dual) perspectivities gives us the full flexibility of projectivities: transformations that preserve incidence and the cross-ratio of four points, as will be discussed in Chapter 6. In mathematical notation, projectivities are often written as: $A(x)\barwedge B(y)$ or $a(X)\barwedge b(\mathrm{Y})$, where the hat over the dash indicates a general projectivity between points (or lines), possibly composed of multiple perspectivities. It can be shown that every projectivity is a composition of at most three perspectivities. Figure 4.9 shows a projectivity generated by two perspectivities with axis m and n. In mathematical notation, this can be expressed as:

$$A(x)\overset{m}{\underset{\wedge}{=}}B(y)\overset{n}{\underset{\wedge}{=}}C(z),\text{ or more concisely, as }A(x)\barwedge C(z).$$

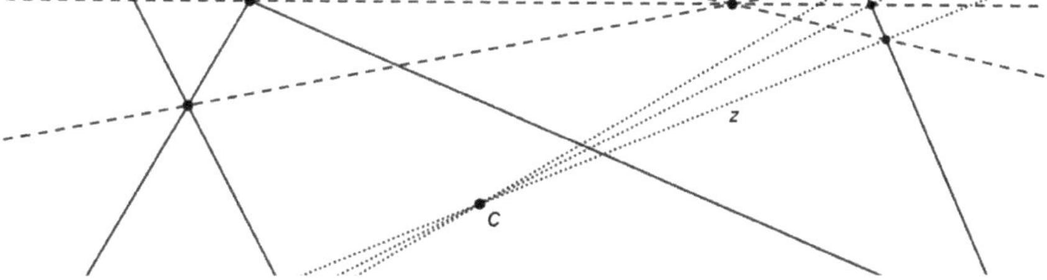

Figure 4.9. A projectivity generated by two perspectivities.

An *autoprojectivity* is a projectivity that maps points and lines of a projective space onto themselves.

In the projective plane, projectivities can be classified based on their fixed elements. For example: a *perspectivity* fixes a point and a line; a *homology* fixes a point (the *center*) not lying on a fixed line (the *axis*); an *elation* fixes both a point on the axis and the axis itself; an *involution* swaps pairs of points; and a general *projectivity* may have no fixed elements at all.

To illustrate this, consider a projectivity π that fixes a line a (the axis) and a point C not lying on that line (the center). Such a transformation is called a *homology,* as shown in Figure 4.10. For instance, a point P that is not on the axis is mapped to another point $\pi(P)$, determined by the geometry of the configuration. If the fixed-point C lies *on* the axis a instead, the transformation is called an *elation*.

Another important type is the *involution* — a projectivity that swaps pairs of points on a line. Depending on the configuration, it may have two fixed points, one, or none at all. By the principle of duality, there is also a dual notion: just as an involution exchanges point pairs, its dual exchanges pairs of lines through a point. In this setting, some lines may remain fixed (i.e., pass through the same point before and

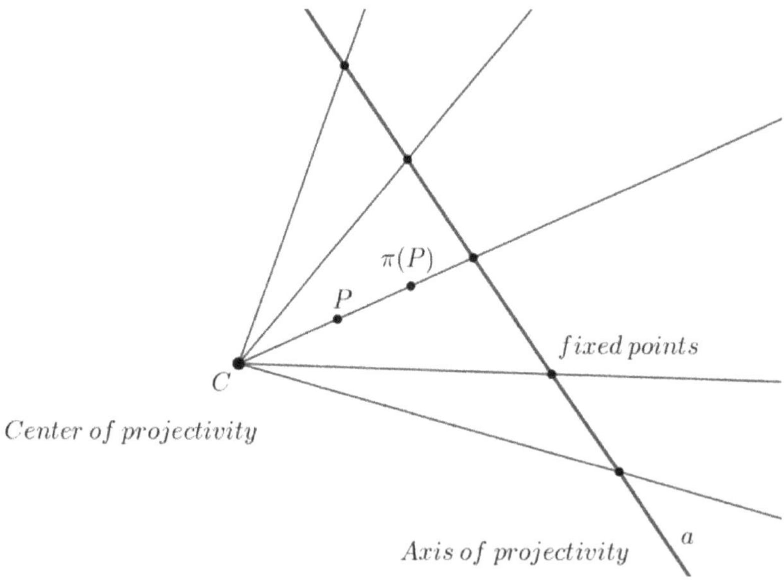

Figure 4.10. A homology.

after the transformation), while others are interchanged. The dual involution may likewise have two, one, or no fixed lines.

We will not go into technical details here, as this would go beyond the scope of this book.

We will explore how projectivities can be used to generate conic sections in Chapter 8, and we will study autoprojectivities and the problem of finding their fixed elements in Chapter 5, where we will need them in a concrete application.

4.5 Further Fundamental Axioms of Projective Geometry

When working seriously with projective geometry and aiming to go deeper — for example, to prove theorems — a few additional axioms become essential.

Although we will not study complicated proofs in this book, we still want to gain insight into some of the key ideas of projective geometry. For that reason, the most important of these axioms will be briefly introduced here, along with an explanation of what they mean.

These include:

- *Desargues' theorem*, which states that if two triangles are in perspective from a point, then the intersections of their corresponding sides lie on a straight line,
- *Pappus's theorem*, which states that given two lines with three points each, the intersections of certain line pairs lie on a common line,
- *The Fano Axiom*, which states that the three diagonal points of a complete quadrangle are *not* collinear — a condition violated by the smallest finite projective plane (see Chapter 4.1), and
- *Pascal's theorem*, a classical result about six points on a conic, which serves as a bridge between projective geometry and the theory of conic sections.

Each of these statements reveals a deeper structural property of projective geometry.

Desargues' Theorem

Girard Desargues (1591–1661) was a French mathematician, engineer, and architect whose pioneering work laid the foundations of projective geometry. He is best known for Desargues' theorem, which states:

> *"If two triangles are in perspective from a point, then they are in perspective from a line."*

To make it clearer: Let two triangles in Figure 4.10, ΔABC and $\Delta A'B'C'$, be such that the lines AA', BB' and CC' all meet at a single point (called the center of perspectivity). Then, the points of intersection of the corresponding sides — $X = AB \cap A'B'$, $Y = BC \cap B'C'$ and $Z = CA \cap C'A$ — lie on a common line, called the axis of perspectivity.

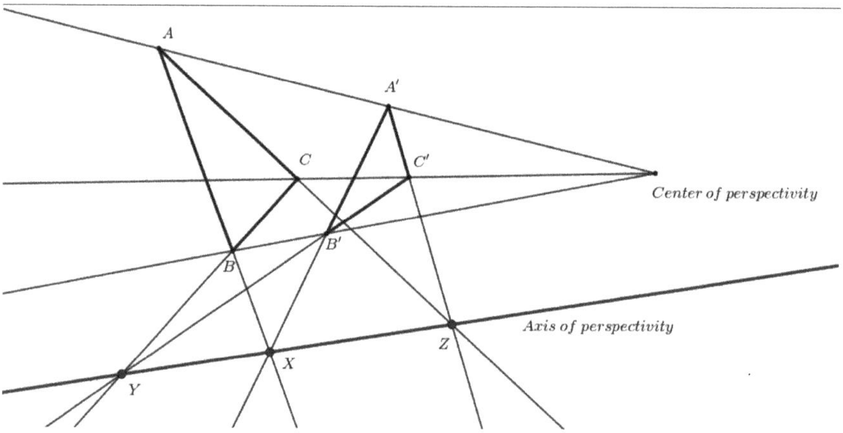

Figure 4.11. Desargues' Theorem.

Naturally, the dual version of the theorem also holds true. To do this, you simply have to read Fig. 4.11 in the other direction — starting with the collinear points X, Y and Z on the line (the axis) and observing that the connections between the corresponding triangle corners then all three meet at a single point (the center).

The projectively completed Euclidean plane is Desarguesian, meaning that whenever two triangles are in perspective from a point, they are also in perspective from a line — and the dual statement holds as well, that is, if two triangles are in perspective from a line, then they are in perspective from a point.

Do non-Desarguesian planes actually exist? Yes, they do. These are projective planes where Desargues' theorem does not hold true. One famous example is the Moulton plane, named after Forest Ray Moulton (1872–1952), an American mathematician and astronomer. It is a modified version of the projectively completed Euclidean plane, where (only) lines with negative slope bend at the y-axis, as can be seen in Figure 4.12. Although it satisfies the basic axioms of incidence geometry (A1)–(A3) (see Chapter 4.1), it is not Desarguesian.

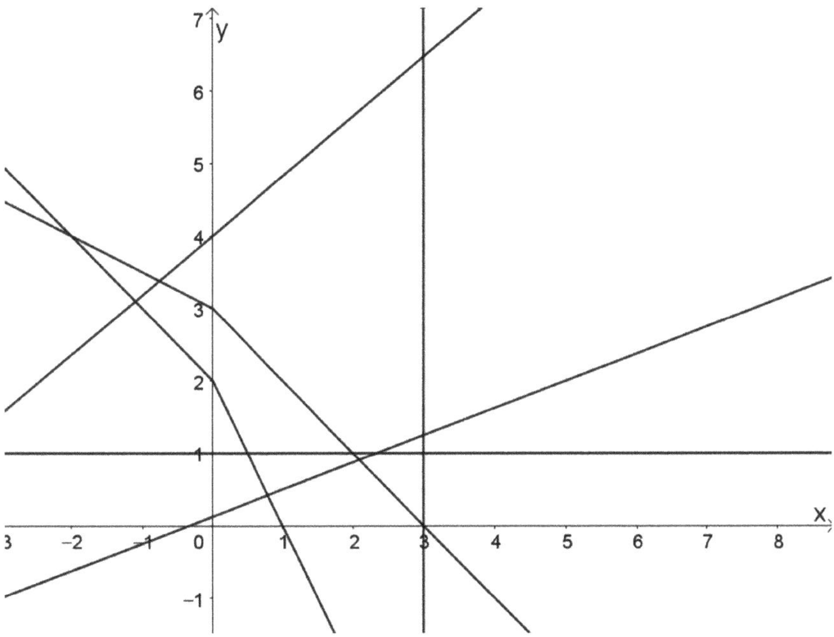

Figure 4.12. Moulton plane.

To be more precise, the Moulton plane is a modified version of the projectively completed Euclidean plane in which certain lines are altered in a specific way. Normally, straight lines are described by the familiar equation $y = mx + b$, where m represents the slope and

b the y-intercept. In the Moulton plane, however, this rule changes — but only for lines with negative slope ($m < 0$). When such a line reaches the y-axis (i.e., when $x = 0$), it "bends" sharply. To the left of the y-axis, the line follows its usual slope m, but to the right, it continues with a steeper slope, typically $2m$. This results in a visible "kink" at the y-axis.

Although the Moulton plane satisfies the basic incidence axioms (A1)–(A3), it does not satisfy Desargues' theorem. To give the reader an idea — without going too far into technical details — a configuration involving two congruent right triangles can illustrate the failure of Desargues' theorem in the Moulton plane, as shown in Figure 4.13. If the triangles are placed on opposite sides of the y-axis, with legs parallel to the coordinate axes, they are perspective from a point at infinity F (see Chapter 4.2). However, the intersections of corresponding sides are not collinear: two lie at infinity (X and Y), while the third (Z) is a finite point resulting from the bending of lines at the y-axis.

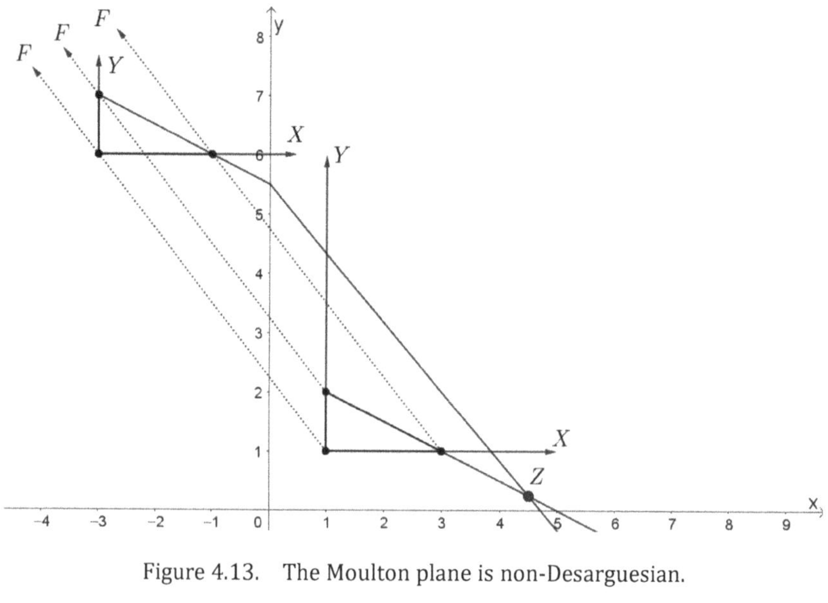

Figure 4.13. The Moulton plane is non-Desarguesian.

The Moulton plane serves as a striking example that Desargues' theorem is not a logical consequence of the basic incidence axioms. Interested readers are encouraged to consult the earlier cited literature for further details.

Pappus's Theorem

Pappus of Alexandria (fl. 290–350 BCE) was a Greek mathematician whose work preserved and extended classical geometry; he is best known in projective geometry for Pappus's theorem, which states:

> *"If six points lie alternately on two lines, then the three intersection points formed by connecting opposite pairs lie on a straight line."*

To make it clearer, in Figure 4.14, let points A, B, C lie on one line and points A', B', C' on another. Then, if we form the intersection points $X = AB' \cap A'B$, $Y = AC' \cap A'C$ and $Z = BC' \cap B'C$ these three points X, Y, Z will always lie on a single straight line — this is the essence of Pappus's theorem.

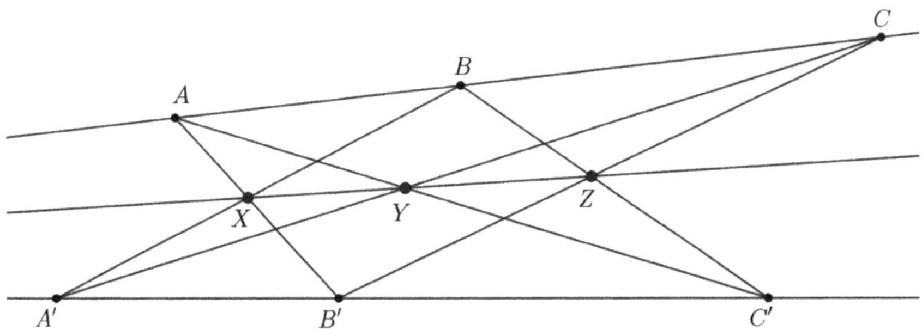

Figure 4.14. Pappus's Theorem.

If Pappus's theorem holds in a projective plane, then its dual version also holds true, that is, in Figure 4.15, if three lines a, b, c pass through a point P, and three lines a', b', c' pass through another point P', then the lines x, y and z connecting the intersection points $a \cap b'$, $b \cap c'$, $c \cap a'$ with their counterparts $a' \cap b$, $b' \cap c$, $c' \cap a$, respectively, all meet in a single point Z — showing that the theorem remains true under the interchange of points and lines, as guaranteed by the principle of duality.

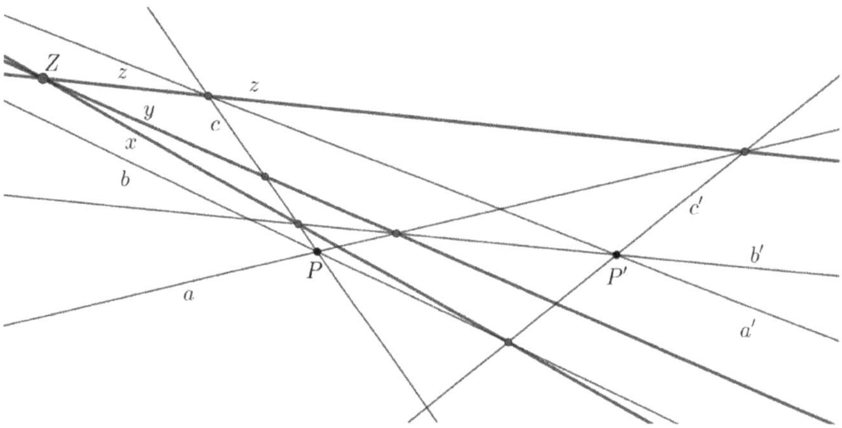

Figure 4.15. Dual Pappus's Theorem.

In the projectively completed Euclidean plane, Pappus's theorem always holds true. If you select six points, three on one line and three on another, and connect them as described in the above way, the resulting three intersection points will always lie on a straight line. But in some exotic geometric systems — called non-Pappian planes — this neat alignment breaks down. These strange planes do not follow all the usual rules, often because their underlying number systems do not behave like regular numbers (for example, multiplication might not be commutative, meaning that something like 3×4 is not necessarily the same as 4×3). While we cannot draw a non-Pappian plane with pen and paper, mathematicians can define and study them using algebra, revealing a world where even basic geometric truths can suddenly fail.

The Fano Axiom

Gino Fano (1871–1952) was an Italian mathematician known for his foundational work in projective and finite geometry; he is best known in this context for the Fano axiom, which states:

> *"The three diagonal points of a complete quadrangle are never collinear."*[1]

In other words, as shown in Figure 4.16, if you draw all the lines connecting four points *A, B, C* and *D* (with no three lying on the same line), then the three points *X, Y* and *Z,* where the opposite sides cross will never lie on a single line.

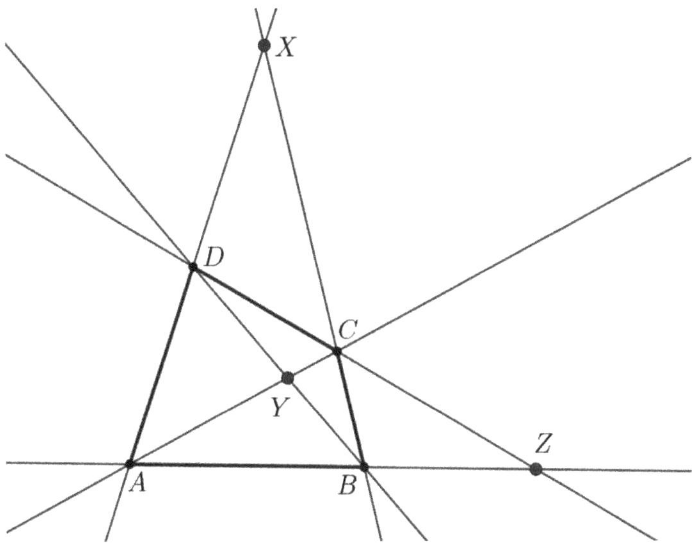

Figure 4.16. The Fano Axiom.

---

[1] A complete quadrangle consists of four points, no three of which are collinear (i.e., no three lie on the same line). Each pair of points determines a line, resulting in six distinct lines connecting the points.

The dual Fano axiom states that the three diagonal lines of a complete quadrilateral[2] never meet at a single point, as seen in Figure 4.17.

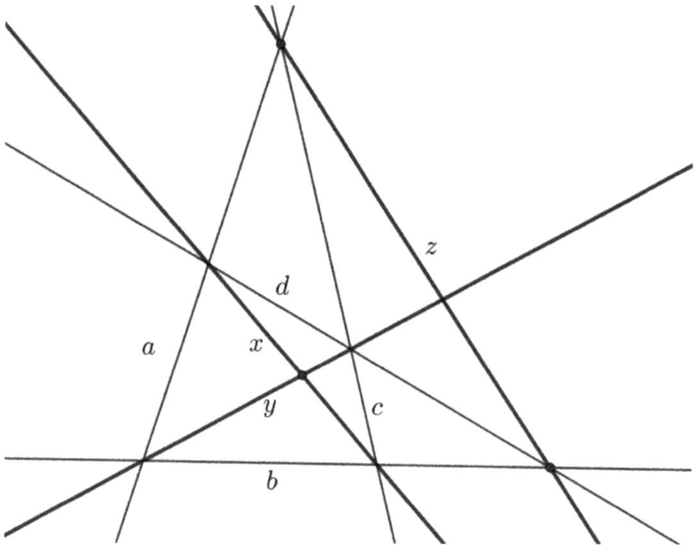

Figure 4.17. Dual Fano Axiom.

The projectively completed Euclidean plane satisfies the Fano axiom — it is Fanoian — which means that in any configuration of four points (no three colinear), the three diagonal points formed by intersecting opposite sides will never lie on the same line.

The minimal projective plane introduced earlier in Chapter 4.1 — also called the Fano plane — is not Fanoian. This means it does not satisfy the Fano axiom: the three diagonal points of a complete quadrangle can lie on a single line. Elementarily, this happens because in the Fano plane, every line contains exactly three points, so whenever you construct a complete quadrangle from four points, there are only seven

[2] A complete quadrilateral consists of four lines, no three of which are concurrent (i.e., no three meet at a single point). These lines intersect in pairs, forming six distinct points. The segments connecting these points are considered the sides of the quadrilateral.

points in total — and some of the diagonal points must fall on the same line, simply because there are not enough points or lines to avoid it (see Figure 4.1).

Pascal's Theorem

Blaise Pascal (1623–1662) was a French mathematician and philosopher whose early work in geometry laid the foundations for projective geometry; he is best known in this context for Pascal's theorem, which states:

> *"If six points lie on a conic section (such as a circle or ellipse), then the three intersection points formed by extending opposite sides of the resulting hexagon lie on a straight line."*

In other words, in Figure 4.18, if you place six points *A, B, C, D, E* and *F* on a conic and connect them in order to form a hexagon, then extend the opposite sides — *AB* and *DE*, *BC* and *EF*, and CD and FA — the three points *X, Y* and *Z*, where these pairs of lines intersect, will always lie on a single straight line.

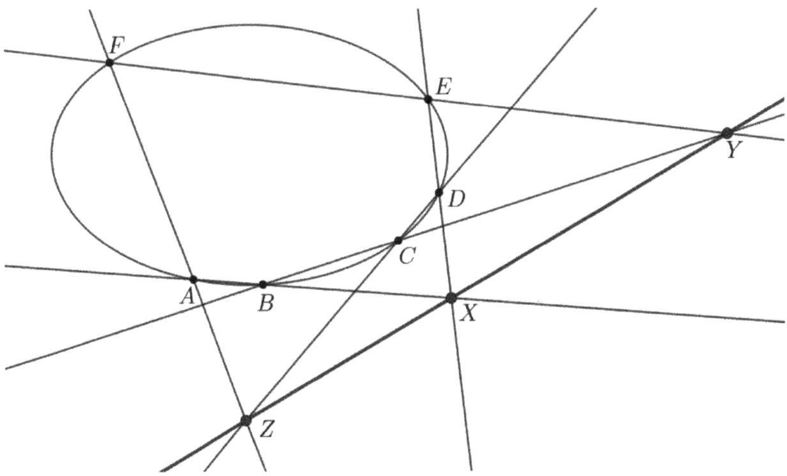

Figure 4.18. Pascal's Theorem.

The dual of Pascal's theorem goes by a different name: Brianchon's theorem. While Pascal's theorem concerns six points on a conic and the collinearity of certain intersection points, Brianchon's theorem deals with six lines tangent to a conic and the concurrency of diagonals formed by connecting opposite vertices of the hexagon.

Charles Julien Brianchon (1783–1864) was a French mathematician and military officer, best known in projective geometry for this result. Brianchon's theorem states:

> *"If six lines form a hexagon that is circumscribed around a conic section, then the three main diagonals (each connecting opposite vertices) meet in a single point."* (See Figure 4.19.)

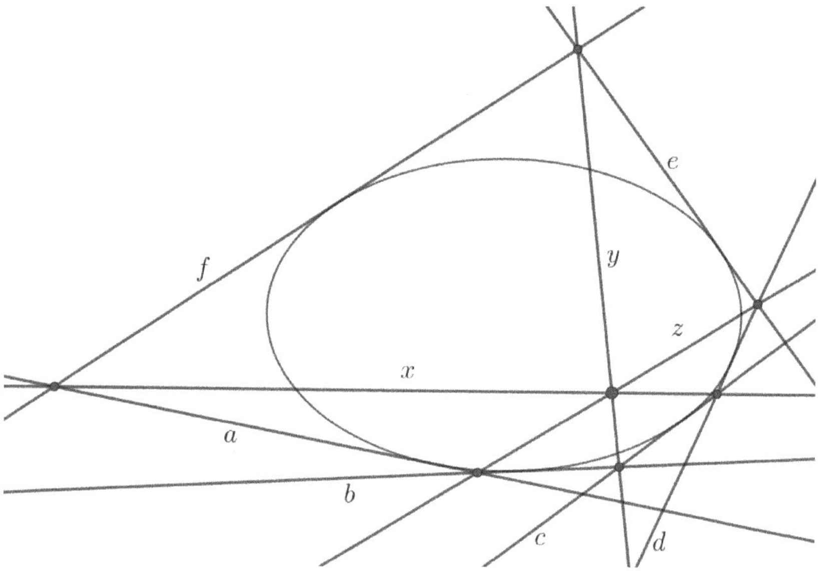

Figure 4.19. Brianchon's Theorem.

Pascal's theorem and its dual, Brianchon's theorem, hold in the projectively completed Euclidean plane. However, they are not valid in all incidence geometries. For instance, the Moulton plane (see above) satisfies the basic incidence axioms (A1)–(A3), but fails to satisfy both Pascal's and Brianchon's theorem. This shows that these results, like Desargues' theorem, are not purely consequences of the incidence axioms, but rely on additional geometric or algebraic structure.

In this chapter, we have explored the key tools of planar projective geometry, focusing on an intuitive, synthetic approach rooted in axioms and incidences rather than algebraic formulations or calculations. It is worth noting, however, that projective geometry can also be approached from entirely different, more abstract algebraic perspectives — for example, through coordinates, vector spaces, or field theory. These alternative frameworks open the door to powerful generalizations, but the geometric insight gained from the classical, axiomatic viewpoint continues to offer a clear and compelling foundation for understanding projective geometry.

If the reader is interested in this algebraic perspective, a classic reference is Semple and Kneebone's *Algebraic Projective Geometry*, which presents a comprehensive development of projective geometry from the viewpoint of coordinate systems and field theory.[3]

[3] Semple, J. G. & Kneebone, G. T.: *Algebraic Projective Geometry*, Oxford University Press, Oxford (1952).

Chapter 5

Surprising Solutions to Geometrical Problems

In this chapter, we will present an example of a surprising solution to an easy-to-understand geometric problem. We will focus on the stunning elegance of the application of a powerful tool from projective geometry and how it can lead, step by step, to the solution of a problem — without going into too much detail as to why it works in this way, as this would go far beyond the scope of this book. The interested reader is referred to the writings of H. S. M. Coxeter[1] and A. Seidenberg[2] for a more intensive study of the subject.

5.1 The Problem

Consider the triangle *ABC* with sides *a*, *b*, and *c*, and inside this triangle another triangle *KLM* (Figure 5.1). The task now is to find a third triangle *XYZ* whose vertices lie on the sides *a*, *b*, and *c* of the triangle *ABC*, and whose sides run through the vertices *K*, *L*, and *M* of the triangle *KLM* (Figure 5.2).

[1] H. S. M. Coxeter: *Projective Geometry*, 2nd Ed., Springer, New York (1987).
[2] A. Seidenberg: *Lectures in Projective Geometry*, Dover Publications, New York (2005).

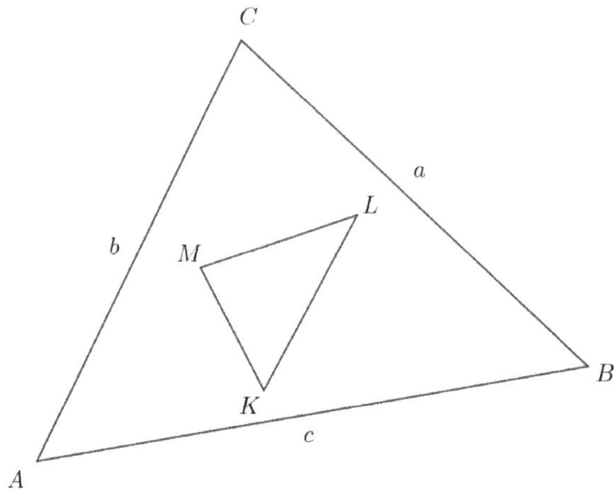

Figure 5.1. Triangles *ABC* and *KLM*.

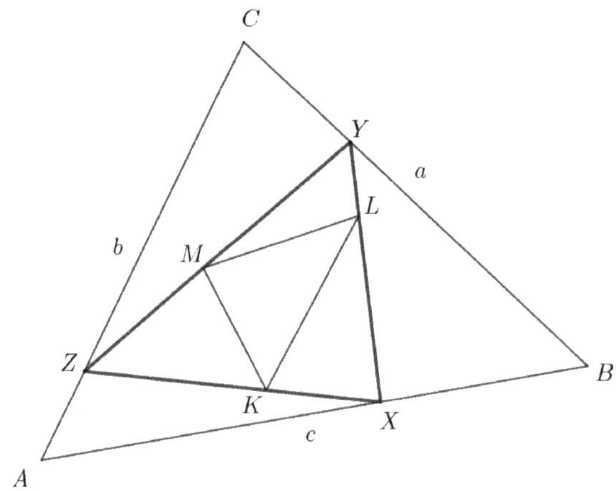

Figure 5.2. Sought-after triangle *XYZ*.

For now, it is not clear whether there always exists such a third triangle *XYZ* to the two given triangles that fulfills the mentioned conditions and if so, whether there can be more than one (and how many). We would very much like to be able to decide for each such pair of triangles ABC and KLM whether there exist solutions (triangles XYZ) to the problem described above, and if so, exactly how many.

5.2 Searching for Solutions

To find a triangle that fulfills the required conditions, we can make a first naive attempt by starting with an arbitrary point X on the side c of triangle ABC, connecting it to K, intersecting the connection with side b — this yields a point Z — then connecting this point Z to M, intersecting the connection with side a — this yields a point Y — and finally connecting point Y to L and intersecting it with side c. When we arrive at starting point X, we have found a solution to the given problem. Unfortunately, this will not to work out in general (Figure 5.3).

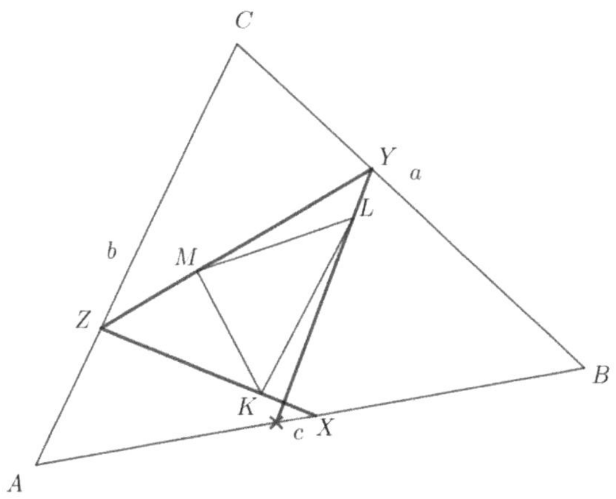

Figure 5.3. Searching for solutions.

5.3 Analysis with Projective Geometry

If we start at an arbitrary point X on the side c and carry out the above chain of constructions, we will generally get a new point X' on c (unless we are lucky enough to have started with a vertex of a solution triangle — then we note this solution and try again to find further solutions). This Point X' is the image of X under a mapping of points on c (which is defined by the above outlined construction). In the language of projective geometry this mapping consists of a chain of perspectivities

$$c(X)\underset{\wedge}{\overset{K}{=}}b(Z)\underset{\wedge}{\overset{M}{=}}a(Y)\underset{\wedge}{\overset{L}{=}}c(X')$$

(see Chapter 4) which leads to a so-called auto-projectivity $c(X)\overline{\wedge}\,c(X')$ on the set of points of c (Figure 5.4).

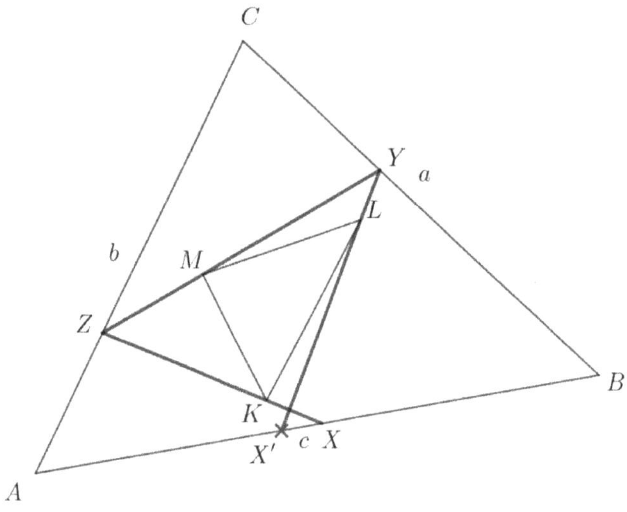

Figure 5.4. Point X and its image X'.

Projective geometry teaches us that such an auto-projectivity (here a projectivity that acts between points on one and the same straight line *c*) always has 0, 1 or 2 fixed points. So, we now know that our problem can have 0, 1 or 2 solutions.

So how can we decide how many solutions our problem has, and how can we find them?

5.4 Using the Steiner Circle for Determining Fixed Points of Auto-Projectivities

Projective geometry allows the determination of fixed points of an auto-projectivity by means of an amazing technique using the Steiner circle: As a first step, we need three pairs of mapped points X_1, X_1', X_2, X_2', and X_3, X_3' (Figure 5.5). We get them by arbitrarily choosing points X_1, X_2, and X_3 on side *c* of the triangle *ABC* and mapping them using the above chain of constructions (thus from perspectivities).

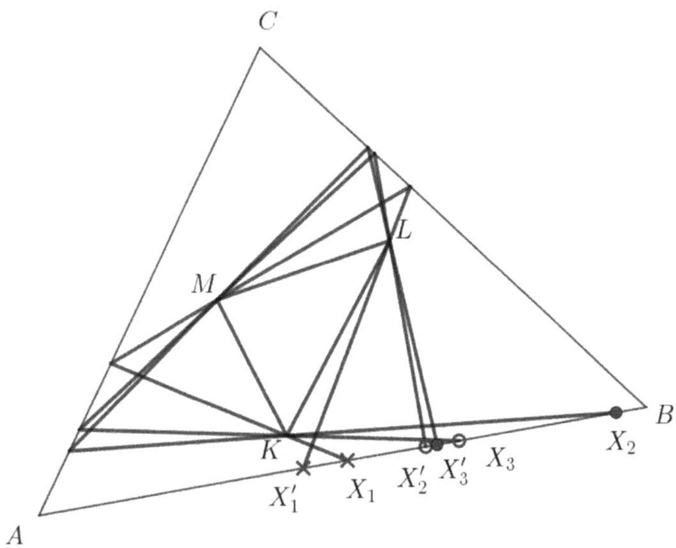

Figure 5.5. Three pairs of points.

Now we can choose any circle s and an arbitrary point S on it (for technical reasons, we choose the point S on s close to our pairs of points on c so that their images on the Steiner circle s are as far apart as possible). We connect the point S with the six points of the three point-pairs X_1, X_1', X_2, X_2', and X_3, X_3' on c and intersect the connecting lines with the circle — so we project these points from point S onto the circle (Figure 5.6). To get a better overview, we will now gradually hide some elements of our previous constructions that we do not need at the moment.

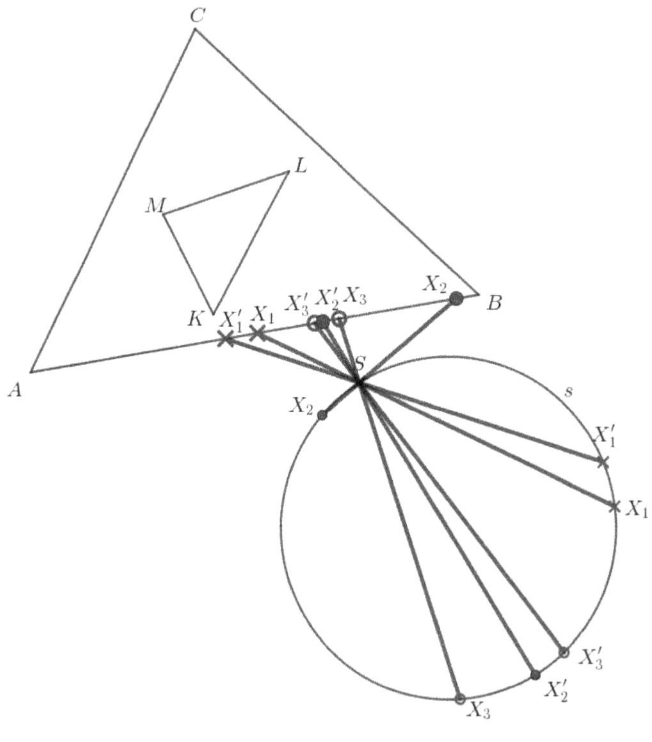

Figure 5.6. Steiner circle *s*.

We can now determine the so-called projectivity axis p by connecting the point pairs on s crosswise and intersecting the connecting lines. For example, we get the point of intersection P_1 by intersecting X_2, X_3' with X_3, X_2' (Figure 5.7).

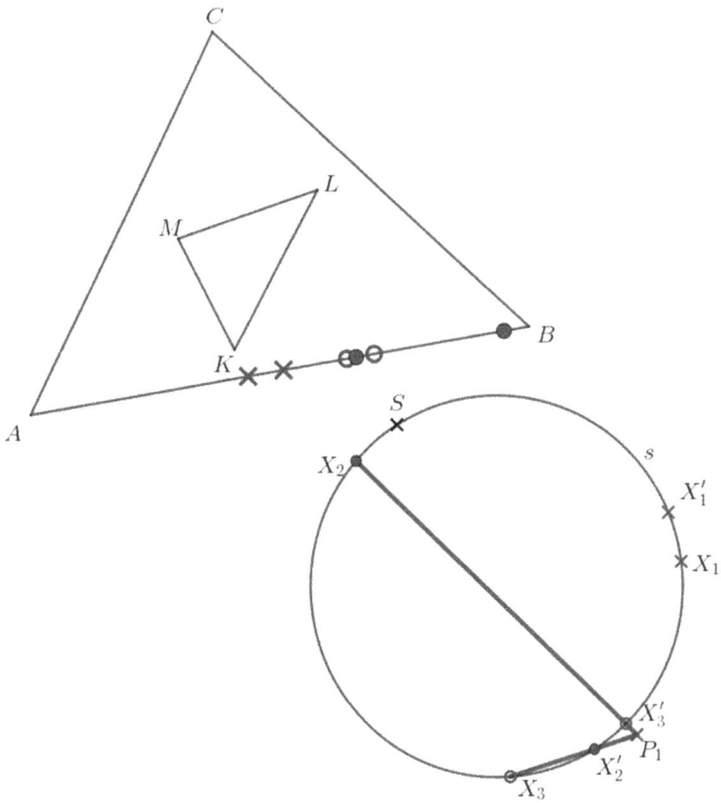

Figure 5.7. Point P_1 of the axis of projectivity p.

If we do this successively for all three pairs of points, we get three collinear points P_1, P_2 and P_3 defining the projectivity axis p (Figure 5.8). Incidentally, it should be mentioned here that the fact that these three points always lie on a straight line is guaranteed by Pascal's theorem (see Seidenberg, p. 110).

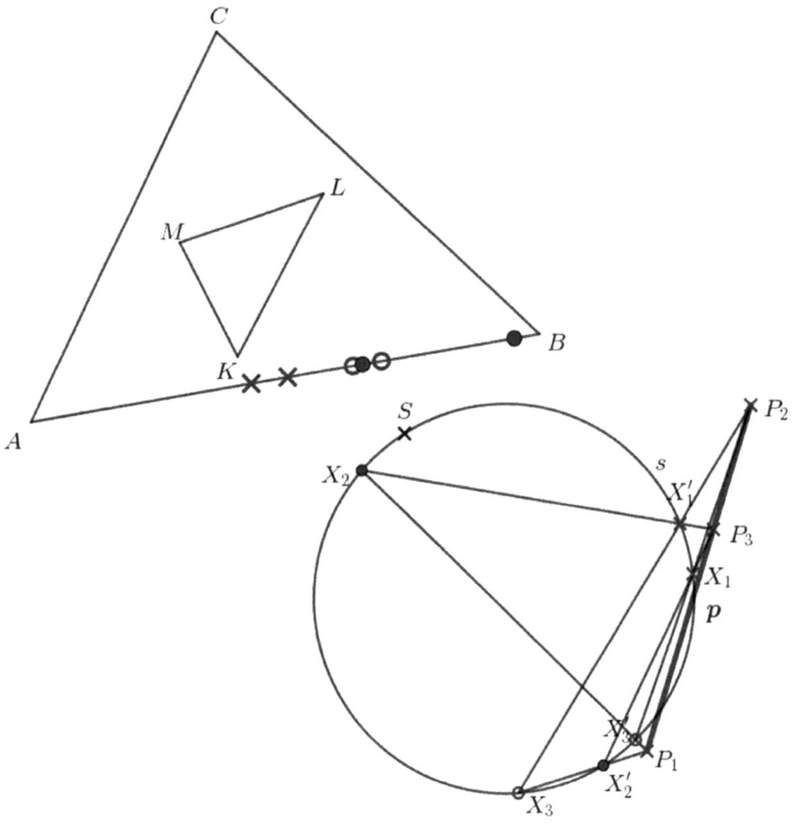

Figure 5.8. Axis of projectivity p.

We see that the axis of projection p has two points of intersection with the Steiner circle s. This tells us, firstly, that the auto-projectivity on c has two fixed points and thus, also our problem has exactly two solutions. Secondly, we obtain these solutions by projecting the two points of intersection S_1 and S_2 from S back onto c (Figure 5.9).

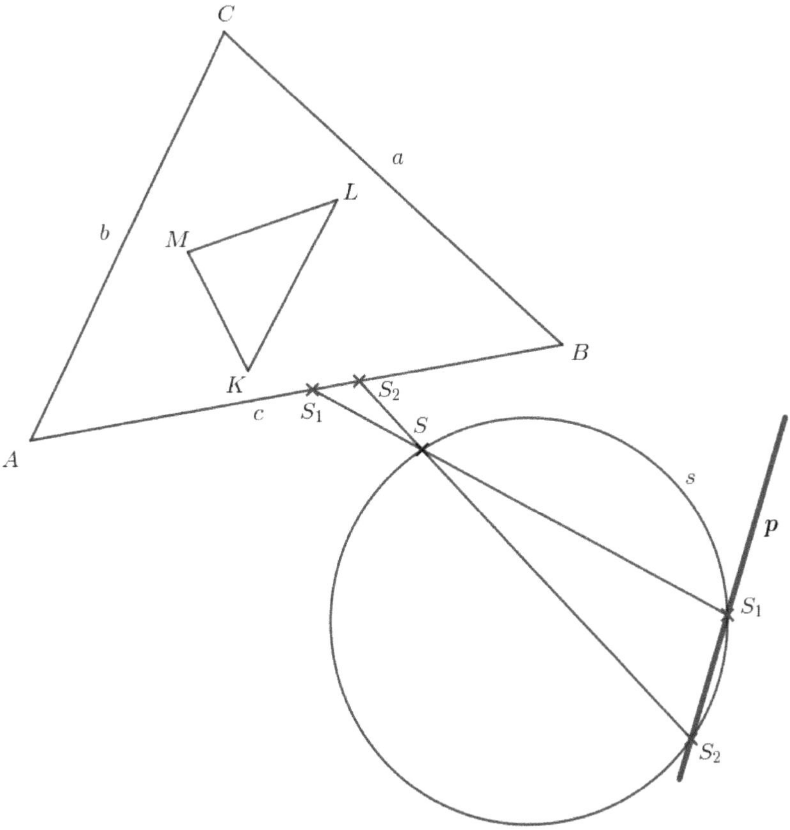

Figure 5.9. Fixed points of the auto-projectivity.

We now want to check whether the two points actually lead to solutions to our problem by executing our construction chain for them again (Figure 5.10).

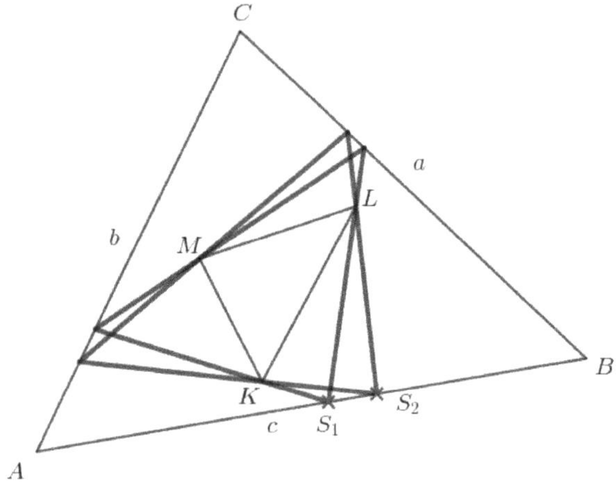

Figure 5.10. The solutions of the problem.

Isn't it amazing that we can find all the exact solutions to our problem by just trying out the construction of any three points on side *c*?

5.5 The Other Solution Cases

Of course, it can also happen that the auto-projectivity has no fixed points, i.e., that the axis of projection has no intersection points with the Steiner circle, and therefore there exists no solution-triangle to the two initial triangles (Figure 5.11).

Although the initial triangles do not look much different from the ones above, there is no solution here, and we can be quite sure of that (since the projectivity axis does not intersect the Steiner circle). We can see what a wonderful tool Jacob Steiner has placed in our hands with his circle!

In order to find starting triangles *ABC* and *KLM* for which there exists exactly one solution, triangle *XYZ*, you would have to construct backwards — starting from a projectivity axis *p* tangential to the

Steiner circle s — since the probability of catching such a constellation by chance is as good as zero.

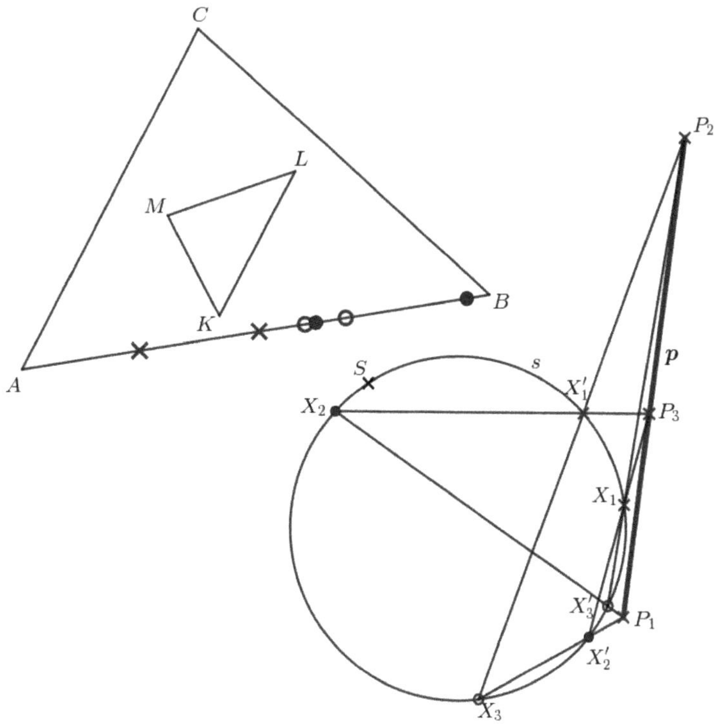

Figure 5.11. No solutions to the problem.

5.6 A Similar Problem

The following problem sounds somewhat similar to what we discussed above:

> *Given are a quadrilateral ABCD and a straight line g on which four points P, Q, R and S lie in different pairs. We are looking for a quadrilateral whose sides pass sequentially through P, Q, R and S and whose vertices lie sequentially on the sides of the quadrilateral ABCD.*

Let's take a closer look at the problem using a sketch (Figure 5.12).

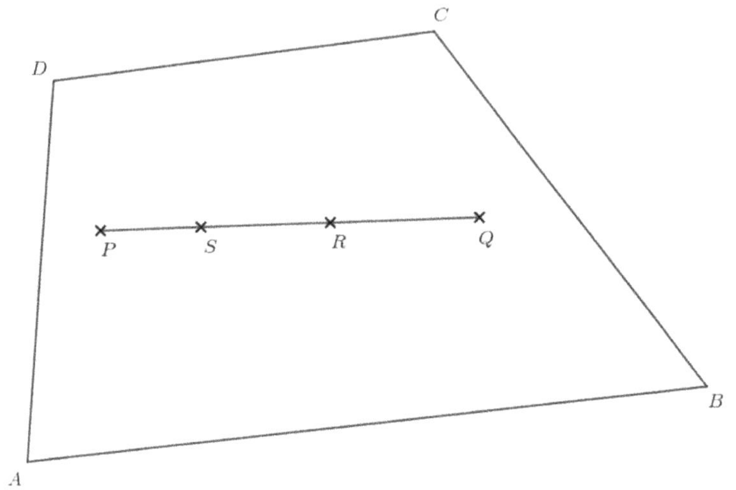

Figure 5.12. The new problem.

This problem strongly reminds us of our problem from above. We can actually solve it the same way. First, we try to find such a quadrilateral triangle that fulfills the required conditions by making the following attempt: We start with an arbitrary point X on AD, connect it with P, intersect the connection with side AB, then connect the point Y obtained with Q, intersect the connection with side BC, then connect the point obtained Y' with R, intersect the connection with side CD and finally connect this point Y'' with S and intersect it with side AD. If we return to starting point X, we have found a solution to the given problem. Unfortunately, as we have already seen regarding the problem above, this will not usually work (Figure 5.13). Again, this whole procedure leads to a chain of perspectivities,

$$AD(X)\underset{\wedge}{\overset{P}{=}}AB(Y)\underset{\wedge}{\overset{Q}{=}}BC(Y')\underset{\wedge}{\overset{R}{=}}CD(Y'')\underset{\wedge}{\overset{S}{=}}AD(X'),$$

which results in a so-called auto-projectivity $AD(X)\overline{\wedge}AD(X')$ on the set of points of AD.

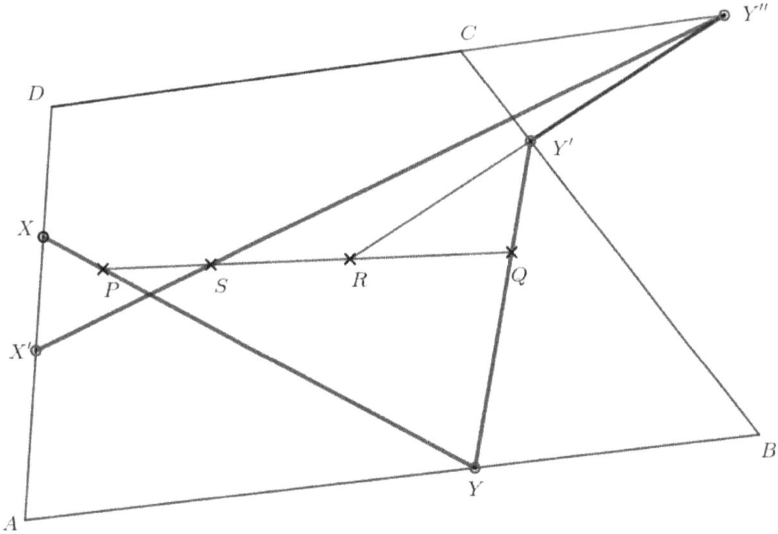

Figure 5.13. Searching for solutions.

Let us just do everything as above. First, we need three pairs of mapped points X_1, X_1', X_2, X_2' and X_3, X_3' (Figure 5.14). We get them by arbitrarily choosing points X_1, X_2 and X_3 on side AD of the quadrilateral $ABCD$ and mapping them using the above chain of constructions.

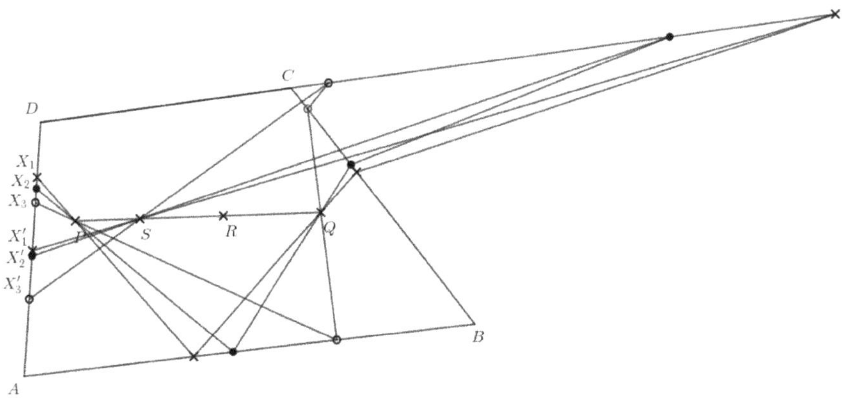

Figure 5.14. Three pairs of points.

Now we can — as above — choose any circle s and an arbitrary point S on it. We connect the point S with the six points of the three point-pairs X_1, X_1', X_2, X_2', and X_3, X_3' on AD and intersect the connecting lines with the circle. We can now determine the projectivity axis p by connecting the point pairs on s crosswise and intersecting the connecting lines (Figure 5.15). We see that the projectivity axis p has two points of intersection with the Steiner circle s. This tells us that our problem has exactly two solutions. We obtain these solutions by projecting the two points of intersection S_1 and S_2 back onto AD (Figure 5.16).

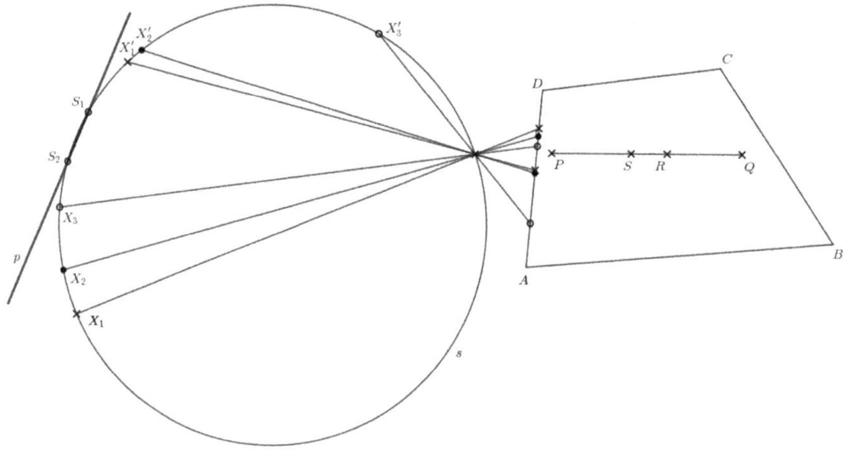

Figure 5.15. Steiner circle, axis of projectivity, and two points of intersection S_1 and S_2.

We can easily convince ourselves that the two solutions are correct, although unfortunately the solution quadrilaterals do not seem very attractive for the given particular specification of points but that's often the case when geometry or math meet reality, not everything always works out well on the drawing sheet and the numbers aren't always pretty either (one of these quadrilaterals is shown in Figure 5.16).

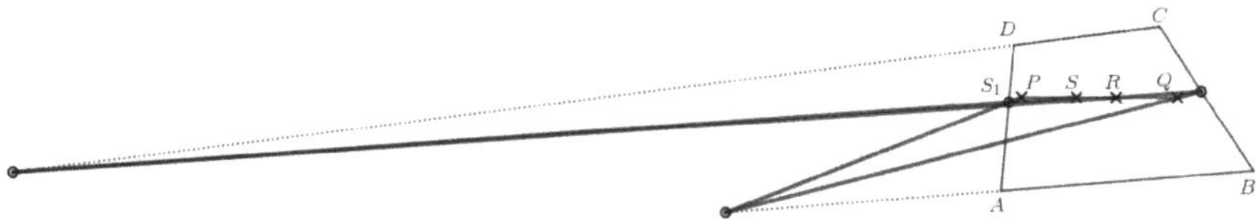

Figure 5.16. One of the two quadrilaterals that solve the problem.

There are many other similar, and less similar, problems that can be solved in the way described above. The amazing thing about the method presented is that you can find the exact solutions of a given problem (and additional knowledge of exactly how many solutions there are) by seemingly pure trial and error (three times). This method is sufficiently demonstrated by the two problems shown. However, going into more depth here would go beyond the scope of this book.

Chapter 6

The Cross-Ratio and Its Application in Locating Midpoints

In Euclidean geometry, distances and angles are fundamental — they allow us to measure how far apart points are or how two lines relate through rotation. These quantities remain unchanged under motions like translation, rotation, or reflection. However, in projective geometry, these notions lose their meaning. When we view the world through perspective — as in photography or drawing — familiar concepts like distance and angle start to behave differently. Distances stretch, angles become distorted, and even parallel lines — like railroad tracks — appear to meet at a point on the horizon. This point is called the *vanishing point*, or in projective geometry, the *point at infinity*. In this perspective, distance and angle lose their usual meaning because they are not preserved under projection. Instead, projective geometry focuses on the properties that remain unchanged, such as whether points lie on the same line (collinearity), whether lines intersect at a point (incidence), and especially the cross-ratio — a special value that stays constant even when the image is heavily distorted, which we will explore more closely in this chapter.

6.1 From Euclidean Ratios to Projective Invariants

Consider 3 points on a straight line — call them A, B, and C. If C lies between A and B, we can ask: how far is C from A, compared to how far it is from B?

We do not even need exact distances. All we care about is the ratio of the two segments: AC to CB. For example, if C is exactly in the middle, the ratio is $1:1$, as shown in Figure 6.1. If C is twice as close to A as it is to B, the ratio might be $1:2$, as shown in Figure 6.2. This kind of ratio — how one point divides a segment between two others — is called a *division ratio*.

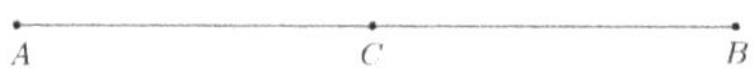

Figure 6.1. The ratio of the two segments is 1:1.

Figure 6.2. The ratio of the two segments is 1:2.

But what happens if the point C lies outside the segment instead of between the two endpoints? In that case, we still compute the division ratio using the oriented distances — meaning we allow negative values depending on the direction along the line. For example, suppose point C lies on the extension of segment AB, beyond point B. Then the length CB becomes negative, because we are moving "backwards" along the line. As a result, the ratio $AC:CB$ becomes negative as well. This negative value tells us that point C no longer lies between A and B, but instead point C divides the line in the opposite direction.

Figure 6.3. Negative division ratio.

In Figure 6.3, you can see such a case. Although the geometry is just as simple as before, the sign of the ratio now carries information about the configuration of the points. Figure 6.3 shows a concrete example where the oriented distances are $AC = 3$ and $CB = -2$, resulting in a division ratio of $3:(-2) = -\frac{3}{2}$. Using oriented segment lengths is essential for understanding more advanced concepts like the cross-ratio — because orientation matters.

Now, let us add a fourth point, D. This allows us to ask a deeper question: how does D relate to the way C divides the line between A and B? We can compare their positions by forming two division ratios:

$$(A,B;C)=\frac{AC}{CB} \text{ and } (A,B;D)=\frac{AD}{DB}$$

(the semicolon here separates the two fixed points A and B from the third point being considered). Then, we take the ratio of those two ratios. At this point, the *cross-ratio* enters the picture. Mathematically, if the points A, C, D, and B lie on a line (and in that order), the cross-ratio is defined as the ratio of these two ratios:

$$(A,B;C,D)=\frac{AC}{CB}:\frac{AB}{DB}.$$

With a bit of algebra, we can rewrite this expression into a more compact form:

$$(A,B;C,D)=\frac{AC\cdot DB}{AD\cdot CB},$$

where each quantity like XY denotes the oriented (or assigned) distance from point X to point Y. That is, we define $XY=Y-X$, so that moving from X to Y gives a positive value when Y lies to the right of X, and a negative value when it lies to the left (assuming points lie on a number line or coordinate axis). This convention allows us to keep track of direction and is important for the properties of the cross-ratio.

The order of the points matters, as permuting them can change both the value and the sign of the cross-ratio — or transform it into one of its six projectively equivalent forms, as we will see further below.

If both division points C and D lie between A and B — that is, both divide the segment internally — then the cross-ratio is positive. Likewise, if both points lie outside the segment on the same side, the result is again positive. But if one point lies between A and B while the other lies outside, the cross-ratio becomes negative. In this way, the sign of the cross-ratio encodes how the four points are arranged along the line.

Let us look at a concrete numerical example to see how the cross-ratio works in practice. Consider four points on the real line: $A=0$, $C=1$, $D=3$, and $B=4$ as shown in Figure 6.4.

Figure 6.4. The cross-ratio of four points.

To compute the cross-ratio $(A,B;C,D)$, we use the definition introduced above, taking into account the order of the points along the line:

$$(A,B;C,D)=\frac{AC\cdot DB}{AD\cdot CB}=\frac{1\cdot 1}{3\cdot 3}=\frac{1}{9}.$$

This simple configuration illustrates how the cross-ratio is calculated from four collinear points using only oriented distances. The cross-ratio is positive, which reflects the fact that both division points C and D lie between the outer points A and B. The sign carries essential geometric meaning, in that, had one of the points been outside that interval, the result would have been negative.

At first glance, the cross-ratio may seem like a simple combination of oriented distances. However, it possesses a remarkable property in that its value remains unchanged under projective transformations — including perspectivities and more general projectivities, which we encountered in Chapter 4. We accept this result from projective geometry without proof, as a full derivation would go beyond the scope of this book. Figure 6.5 shows, as an example, a configuration in which the cross-ratio is preserved under the perspectivity

$$m(X)\overset{L}{\underset{\wedge}{=}}n(Y).$$

Whether the configuration of points is stretched, tilted, or projected from a different viewpoint, the cross-ratio remains invariant. This makes it one of the fundamental tools in projective geometry: a quantity that stays constant even when distances and angles no longer do.

In everyday life, this is the mathematics hiding behind perspective drawing, camera lenses, and even certain visual illusions.

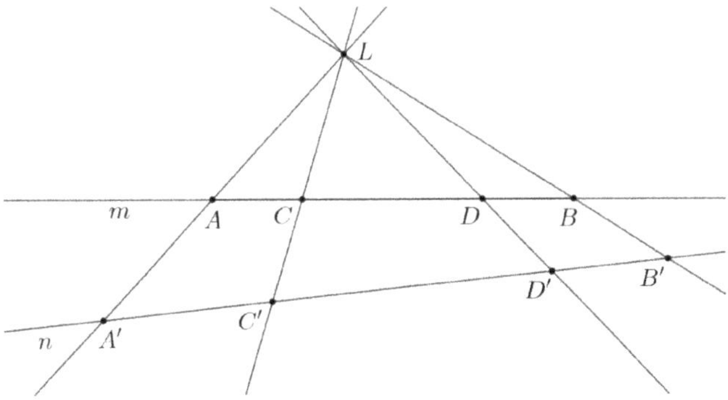

Figure 6.5. $(A,B;C,D)=(A',B';C',D')$.

The following identities show how the value of the cross-ratio changes under different orderings of the four points. These transformations can be verified by direct computation using the definition of the cross-ratio $(A,B;C,D)=\dfrac{AC\cdot DB}{AD\cdot CB}$:

$$(A,B;C,D)=(B,A;D,C)=(C,D;A,B)=(D,C;B,A)=\lambda$$

$$(A,B;D,C)=(B,A;C,D)=(C,D;B,A)=(D,C;A,B)=\frac{1}{\lambda}$$

$$(A,C;B,D)=(B,D;A,C)=(C,A;D,B)=(D,B;C,A)=1-\lambda$$

$$(A,C;D,B)=(B,D;C,A)=(C,A;B,D)=(D,B;A,C)=\frac{1}{1-\lambda}$$

$$(A,D;B,C)=(B,C;A,D)=(C,B;D,A)=(D,A;C,B)=\frac{\lambda-1}{\lambda}$$

$$(A,D;C,B)=(B,C;D,A)=(C,B;A,D)=(D,A;B,C)=\frac{\lambda}{\lambda-1}$$

This covers all $4!=1\cdot2\cdot3\cdot4=24$ permutations of the points *A, B, C,* and *D*. The cross-ratio reduces them to six distinct values, making it a fundamental invariant in projective geometry.

6.2 A Special Case: The Harmonic Quadruple

Among all possible values of the cross-ratio, the that one stands out as particularly important is:

$$(A,B;C,D)=-1.$$

When the cross-ratio of four collinear points equals −1, we say that the points *A, B, C,* and *D* form a *harmonic quadruple.* This configuration has remarkable geometric properties. In particular, point *D* is said to be the harmonic conjugate of point *C* with respect to *A* and *B.* This means that *D* balances *C* in a precise and symmetric way — not in terms of distance, but in the projective sense. Harmonic quadruples appear naturally in constructions involving reflections, angle bisectors, and perspectivities, and they play a central role in classical projective geometry.

Without proof, we accept the following remarkable fact from projective geometry: using a complete quadrangle (a configuration formed by four points in general position — meaning no three lie on a line — and the six lines connecting them), we can construct four collinear points that form a harmonic quadruple, meaning their cross-ratio equals −1, as shown in Figure 6.6.

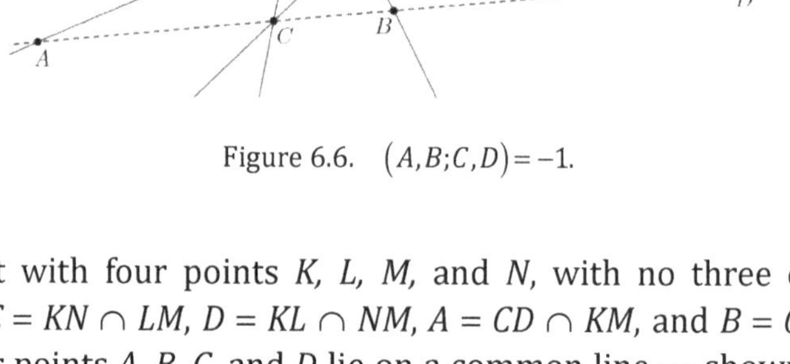

Figure 6.6. $(A,B;C,D)=-1.$

Start with four points *K, L, M,* and *N,* with no three collinear. Define $C = KN \cap LM$, $D = KL \cap NM$, $A = CD \cap KM$, and $B = CD \cap LN$. The four points *A, B, C,* and *D* lie on a common line — shown dashed in Figure 6.6 — and satisfy $(A,B;C,D)=-1$.

If four collinear points A, B, C and D form a harmonic quadruple — that is, if the cross-ratio $(A,B;C,D)=-1$ — then this also holds for the permutation $(A,B;D,C)$ since

$$(A,B;D,C)=\frac{1}{(A,B;C,D)}=\frac{1}{-1}=-1,$$

as we have seen earlier. This makes sense geometrically as well: in a harmonic division, one point lies between A and B, while the other lies outside — and swapping the two simply reverses their roles, without changing the essential balance.

The interested reader may verify, for example, that the harmonic relation $(A,B;C,D)=-1$ holds for the quadrangle defined by the points $K=(0,0), L=(4,1)$, $M=(3,5)$ and $N=(1,4)$, as shown in Figure 6.7, within the projectively extended Euclidean plane. In this configuration, the intersection point C of lines KL and NM has coordinates

$$\left(-14,-\frac{7}{2}\right),$$

and the intersection point D of lines KN and LM is at

$$\left(\frac{17}{8},\frac{17}{2}\right).$$

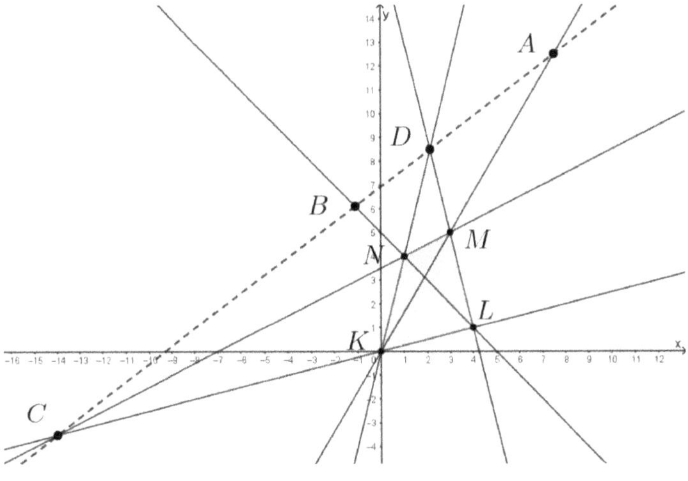

Figure 6.7. $(A,B;C,D)=-1$.

The line through C and D then intersects the lines KM and LN at the points

$$A=\left(\frac{15}{2},\frac{25}{2}\right) \text{ and } B=\left(-\frac{11}{10},\frac{61}{10}\right),$$

respectively. These four points lie on a common line and satisfy the harmonic condition $(A,B;C,D)=-1$. This result can be confirmed by directly calculating the cross-ratio using basic vector methods in the plane — that is, by computing the directed distances between the points and substituting them into the formula

$$(A,B;C,D)=\frac{AC\cdot DB}{AD\cdot CB}=\frac{26.8\cdot 4.02}{6.7\cdot(-16.08)}=-1.$$

Since the cross-ratio remains unchanged under projective transformations, and because any two complete quadrangles (with no three points on a line) can be mapped onto each other by such a transformation of the projectively extended plane, we only need to verify the harmonic condition $(A,B;C,D)=-1$ in one specific example. As a result of this invariance, if the condition holds in one configuration, it must hold in all projectively equivalent configurations. In other words, checking just one well-chosen case — like the one shown in Figure 6.7 — is enough to confirm that the harmonic property is a general geometric truth.

6.3 Finding the Harmonic Conjugate of a Point

Given three distinct collinear points A, B, and C, there exists a unique fourth point D (also on the same line) such that the four points A, B, C, and D form a harmonic quadruple — that is, $(A,B;C,D)=-1$. In this situation, point D is called *the harmonic conjugate* of C with respect to A and B. Geometrically, this means that C and D divide the segment AB in such a way that the division is perfectly balanced — but not in terms of length, rather in terms of cross-ratio. If C divides AB internally in a certain ratio, then D divides it externally in the exact opposite ratio. This special configuration, where one point "undoes" the effect of the other, is what makes it harmonic.

Building on our earlier observations about the complete quadrangle, we can construct such a configuration around three given collinear points A, B, and C in such a way that the fourth harmonic point D arises naturally. The idea is to embed the line containing these points into a full projective figure — a complete quadrangle — whose diagonal intersections determine a harmonic quadruple. In this way, the harmonic conjugate D of C with respect to A and B appears as the result of a purely geometric construction, without the need for measuring distances.

We start with three collinear points A, B, and C, with C lying between A and B, as shown in Figure 6.8. To construct the harmonic conjugate of C with respect to A and B, we begin by drawing two arbitrary lines a and b through point A. These lines will serve as sides of a complete quadrangle and provide the projective framework needed for the construction. Next, we choose an arbitrary point K on line b (distinct from A), and connect it to point B. The line KB will intersect line b at a new point L. We then draw a line from point C to point L, and let it intersect line a at point N. Connecting N to point B, we obtain a new line, which intersects line b at point M. This gives us the complete quadrangle $KLMN$, whose sides and diagonals determine the desired harmonic configuration.

Finally, we construct the diagonal line KM and intersect it with the original line containing A, B, and C. The point of intersection is the harmonic conjugate D of C with respect to A and B.

Since a harmonic quadruple $(A,B;C,D)=-1$ requires the cross-ratio to be negative, we already know — even before constructing D — that the fourth point D must lie outside the segment AB.

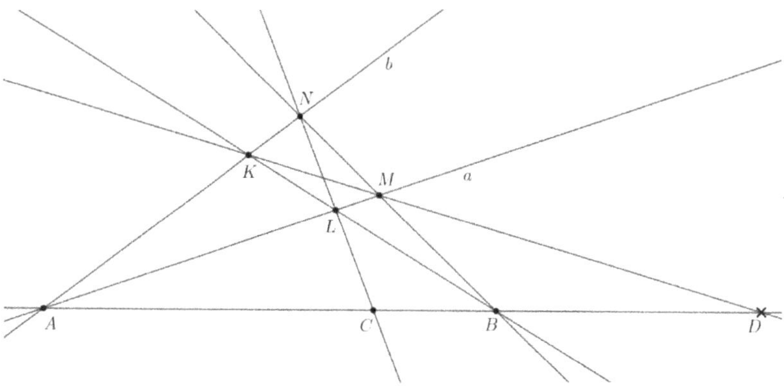

Figure 6.8. Three collinear points A, B, and C.

6.4 Midpoints

Previously, we saw how to construct the fourth harmonic point geometrically, as shown in Figure 6.8. Let us now consider a special case of this construction — namely, when point C is exactly the midpoint of the segment AB. In this situation, the harmonic conjugate D of C with respect to A and B lies on the same line, but outside the segment, dividing it externally in the same ratio. Since C divides AB internally in the ratio 1:1, the point D must divide it externally in the ratio 1:(−1), which characterizes the harmonic condition $(A,B;C,D)=-1$.

When point C is the midpoint of segment AB, and we carry out the standard construction using a complete quadrangle as described above, the diagonal KM and AB turn out to be parallel — as shown in Figure 6.9. This means they do not intersect at a finite point, but at the point at infinity in the direction of the line AB. Hence, in this special case, the harmonic conjugate D of C is the point at infinity on the line — which aligns with the projective interpretation of harmonic division in the ratio 1:(−1).

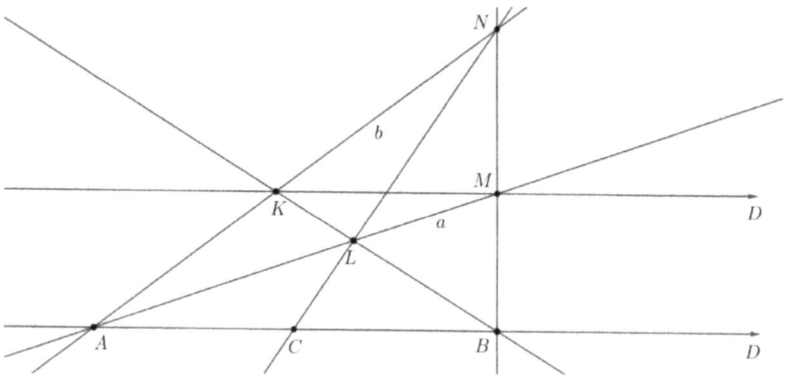

Figure 6.9. Construction of the harmonic conjugate of the midpoint.

This observation leads to an interesting application. Unlike parallel projections — such as orthographic drawings (top views, side views, axonometric projections) — perspective images (like photographs or perspective sketches) do not preserve midpoints. That is, the midpoint of a segment in the image does not necessarily correspond to the actual midpoint in the original object. In addition, points at infinity are generally mapped to vanishing points in a perspective projection. Since perspective images are projective transformations, cross-ratios remain invariant.

This means we can use the cross-ratio to construct the midpoint of a segment in a perspective image. Suppose we know the image points corresponding to A and B, and also the vanishing point D of the line containing them. Then the midpoint C of the segment AB in the original scene can be found as the harmonic conjugate of the vanishing point D with respect to A and B. Using this construction, we can locate true midpoints in perspective images without relying on any metric information — purely through projective geometry. Let us now look at two concrete examples to illustrate this principle.

Figure 6.10 shows a perspective view of a simple house, with the image plane oriented vertically. The vanishing points F_x and F_y, corresponding to the two principal directions of the building, lie on the horizon line h.

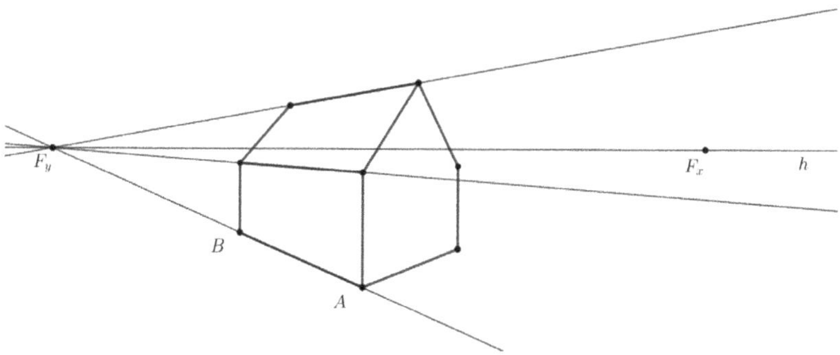

Figure 6.10. Perspective view of a simple house.

We know that F_y is the image of the point at infinity on the line containing segment AB. While the idea of a "point at infinity" may seem abstract — after all, how can you see infinity? — in projective geometry, it stands for the direction in which parallel lines run. In reality, the edges corresponding to segment AB are parallel. But in the drawing in Figure 6.10, they appear to meet at a single point: the *vanishing point* F_y. This is just like railroad tracks that look as if they converge in the distance, even though they remain parallel. In projective terms, this vanishing point represents the point at infinity. Together with the endpoints A and B and the midpoint M, it forms a harmonic quadruple. Since the cross-ratio is preserved under projection, we can determine the image of the midpoint M using the geometric construction shown in Chapter 6.3 — and illustrated here in Figure 6.11.

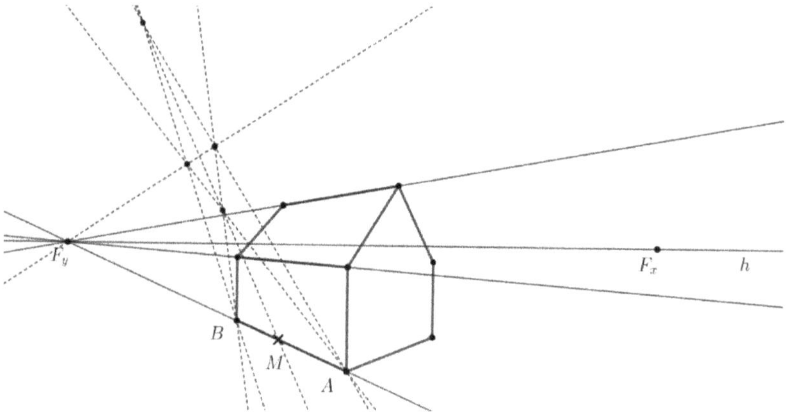

Figure 6.11. Construction of the midpoint.

In a second example, we can examine whether the roof ridge of the house and the central window post of the house shown in Figure 6.12 are aligned with the vertical axis of symmetry in the gable wall. When we construct the midpoint M corresponding to the chosen points A and B — as shown in Figure 6.12 using the dashed lines — we can see that this is not the case, because the point M does not lie on the vertical line passing through the point at the ridge.

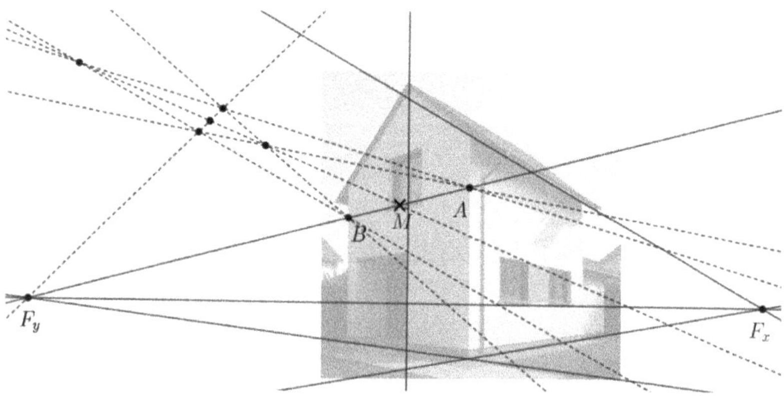

Figure 6.12. Construction of the midpoint.

6.5 Harmonic Point Quadruples and Polarity

In this Section, we will investigate how polarity — a geometric correspondence defined by a circle — leads naturally to involutions, a special class of projective transformations, and harmonic point quadruples. For each point outside the circle, a polar line is defined via the two tangents to the circle, and each line corresponds to a unique pole. This symmetric relationship between points and lines gives rise to involutions on secant lines and naturally leads to harmonic configurations.

While polarity can be defined for all conic sections, we restrict our attention here to the special case of the circle. The general theory for all conic sections will be explored later in Chapter 8. Focusing on the circle allows us to develop the fundamental ideas clearly and geometrically, while already revealing the deep connection between polar constructions and harmonic point sets.

We now introduce the fundamental elements of *pole* and *polar* with respect to a circle. Given a fixed circle c, each point P outside the circle determines a line — the polar p — constructed using the tangents from that point to the circle. This construction illustrates the basic idea of polarity, a central concept in projective geometry. Given a point P outside the circle c, the polar p of point P is the line through the points of tangency T_1 and T_2 of the tangents t_1 and t_2 from point P to the circle c, as shown in Figure 6.13.

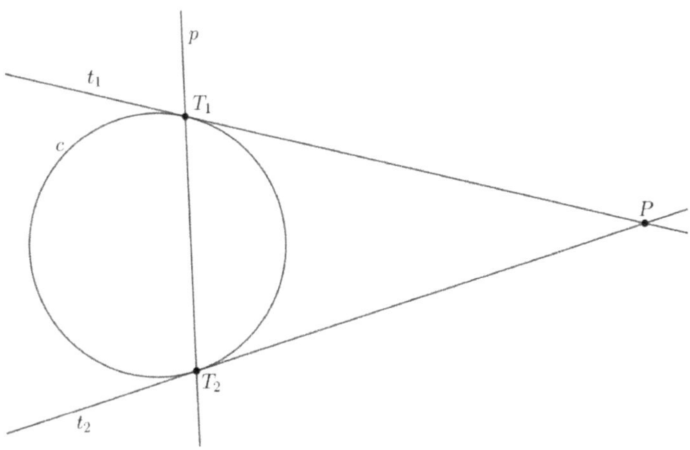

Figure 6.13. Polarity with respect to the circle c.

To further develop the concept of polarity with respect to a circle, we now introduce the inverse point P' of a point P with respect to the circle — not using distances, but purely geometrically, through intersection constructions.

Let P be a point outside a circle c with center O, as shown in Figure 6.14.

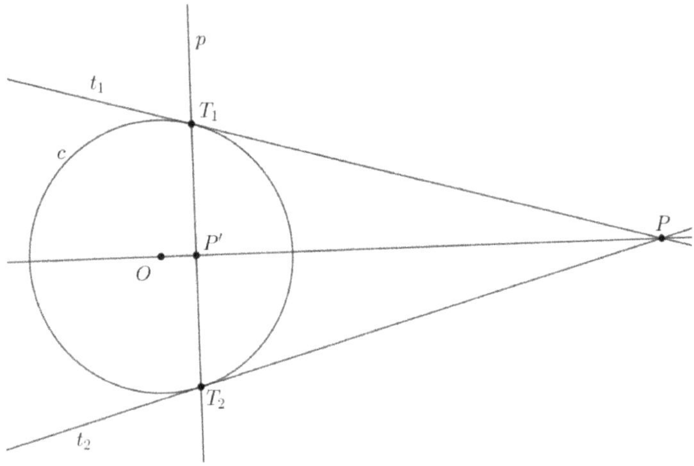

Figure 6.14. Involutive correspondence between P and P'.

The polar p of point P is the line through the two tangency points T_1 and T_2 of the tangents from point P to the circle. Now draw the line through O and P — this is a secant if extended in both directions.

The inverse point P' of P with respect to the circle is then defined as the point where this secant OP intersects the polar p.

An alternative approach to defining the inverse point P' of a point P with respect to a circle is to use the relation $OP \cdot OP' = r^2$, where O is the center and r the radius of the circle. In this formulation, P' lies on the ray OP, and its distance from O is determined algebraically. While this method provides a clear analytic definition, we will not pursue it here. Instead, we focus on the geometric construction based on tangents and intersections. Nevertheless, we mention this distance-based approach for completeness.

Conversely, if we start with a point P' inside the circle, we draw the line OP' and find its two intersection points with the circle. Constructing the tangents at those two points and intersecting them yields a point P outside the circle — the inverse of P'. The line through the two points of tangency is then the polar of P, and by symmetry, also the polar associated with P'.

If we apply this mapping twice — starting from a point P, constructing its inverse P', and then inverting again — we return to the original point P. This kind of behavior is familiar from reflections, where reflecting twice across the same line brings a point back to its original position. This mapping is also known as *circular inversion*, or simply a *circle reflection*.

In projective geometry, any transformation with this property — that applying it twice yields the identity,where every point is mapped back to itself — is called an *involution*. Involutions play a central role in connecting polar geometry, conjugate points, and harmonic point configurations.

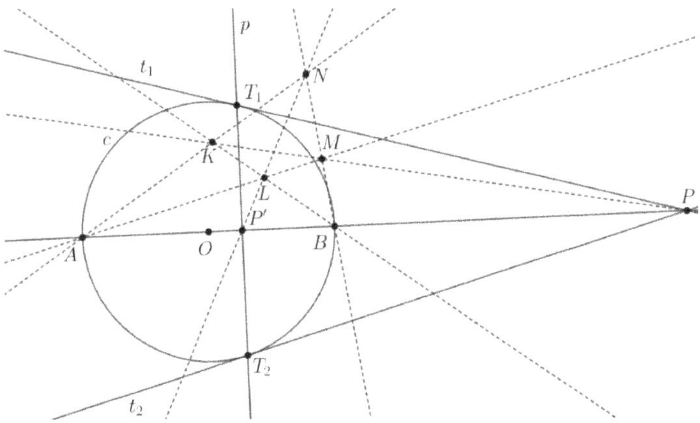

Figure 6.15. Harmonic point quadruple *A, B, P* and *P'*.

If we intersect the line OP with the circle c, we obtain two points A and B. The four points A, B, P, and P' lie in a special configuration: they form a harmonic point quadruple, meaning that their cross-ratio is equal to -1. This can be verified directly using a simple geometric construction: starting from the three points A, B, P, we can construct the harmonic conjugate as described in Section 6.3. Remarkably, this construction leads us precisely to the point P'. The construction — shown in Figure 6.15 with dashed auxiliary lines — illustrates that point P' indeed satisfies the harmonic condition $(A,B;P,P')=-1$.

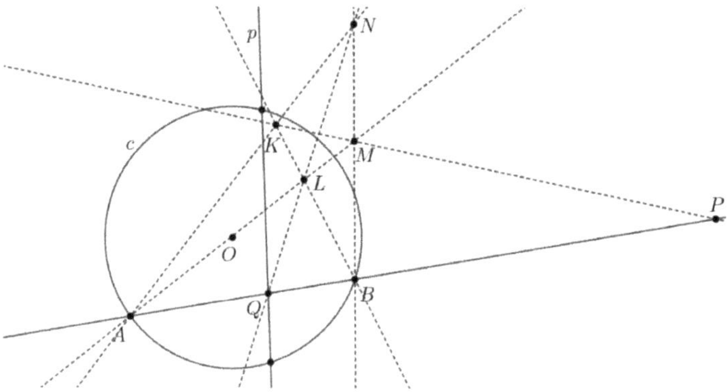

Figure 6.16. Harmonic point quadruple A, B, P and Q.

Surprisingly, this property is not limited to the line OP: as shown in Figure 6.16, it also holds for any secant through point P. If such a line intersects the polar p of P at a point Q, and intersects the circle c at two points A and B, then the four points A, B, P and Q again form a harmonic point quadruple: $(A,B;P,Q)=-1$.

A familiar result from classical Euclidean geometry says that if a point lies outside a circle, and one draws a secant through the circle and a tangent from that same point, then the square of the tangent segment length equals the product of the lengths of the two secant segments. This is known as the *Secant-Tangent Theorem*, and it is a special case of a more general concept: the power of a point with respect to a circle.

The power of a point is defined as the product $PA \cdot PB$, where A and B are the points where any secant through point P intersects the circle, seen in Figure 6.17. This value remains constant for all such secants. In the special case where the line through P is a tangent, both points of intersection coincide, and the power becomes PT^2, where T is the point of tangency.

In projective geometry, this leads naturally to the idea of polarity: the polar line of a point P consists precisely of those points Q that, together with any secant AB through P, form a harmonic point quadruple $(A,B;P,Q)=-1$. In this way, the classical notion of power becomes a bridge between metric geometry and projective ideas like polarity and involution.

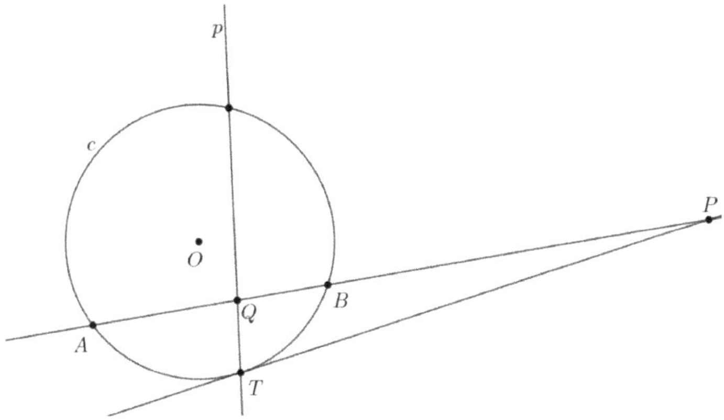

Figure 6.17. Secant-Tangent Theorem.

Now here is the deeper connection: In projective geometry, the polar line of point P with respect to the circle is defined in such a way that for every secant through P, the intersection point Q of this secant with the polar satisfies $(A,B;P,Q)=-1$.

That is, the four points A, B, P and Q form a harmonic point quadruple. The point Q is thus, determined by harmonic division, and the fact that $PA \cdot PB$ is constant ensures that Q always lies on the same line — the polar. In this way, the power of a point (a metric property) and the polar line (a projective concept) describe the same geometric structure from different perspectives.

In the projectively completed plane, we must also consider the special case where the point P lies at infinity — that is, a point at infinity in a given direction. In this case, all secants through point P are parallel lines with that direction, as shown in Figure 6.18. The polar of such a point at infinity is then the line perpendicular to that direction through the center of the circle — that is, the diameter normal to the direction of the secants.

This fits perfectly with the general concept of polarity: The polar of a point at infinity is the line of centers of all circles tangent to lines in that direction. In the case of a circle, this reduces to the diameter orthogonal to the given direction. So even points at infinity have well-defined polars — fully consistent with the projective geometry framework.

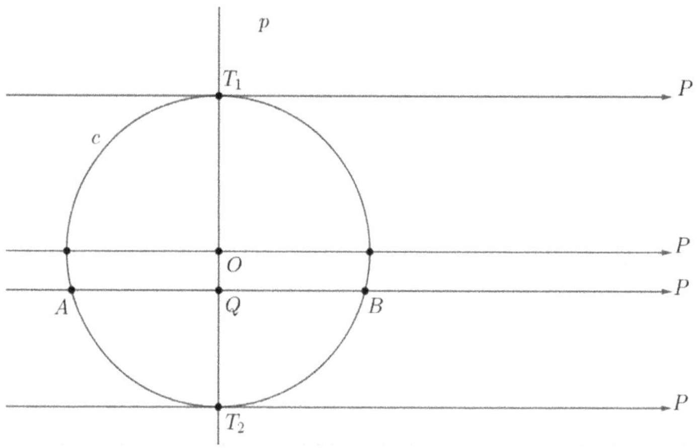

Figure 6.18. *P* is a point at infinity.

On any secant intersecting the circle in two points A and B and drawn in a fixed direction (i.e., corresponding to a point at infinity), the harmonic conjugate of the point at infinity P with respect to A and B is the midpoint Q of the segment AB. That is, the quadruple $(A,B;Q,P)$ is harmonic.

This aligns perfectly with the geometric interpretation of polarity: the polar of the point at infinity in a given direction is the line perpendicular to that direction through M, which — as all such midpoints lie on the same line — is the diameter perpendicular to the direction. The polar and the harmonic structure are, thus, consistent even when one of the points lies at infinity.

In this chapter, we explored the concept of the cross-ratio of four collinear points — a fundamental invariant in projective geometry.

We defined how it is calculated, understood its geometric meaning, and examined the special case of a harmonic point quadruple, where the cross-ratio equals −1. Through examples and constructions, we saw how to determine harmonic conjugates and how the cross-ratio can be used to locate unknown points, such as the midpoint of a segment — even when that segment appears distorted by perspective, as in a photograph.

These ideas demonstrate the strength of projective reasoning: certain relationships remain valid even when shapes are distorted by projection. The cross-ratio is one such relationship — it is preserved under all projective transformations, making it an essential tool far beyond classical geometry.

As an outlook, we have seen that the cross-ratio plays a role in computer graphics and image processing. When analyzing perspective views, the cross-ratio helps with camera calibration, measuring distances, or even reconstructing 3-D scenes from flat images. It is one of the few quantities that remain unchanged under perspective projection.

Finally, the cross-ratio also appears in complex analysis. A special class of functions called *Möbius transformations* maps the complex plane to itself while preserving the cross-ratio. This makes the cross-ratio a kind of "projective fingerprint" — a bridge between geometry, analysis, and modern applications. What began as a construction with four points on a line turns out to be a key idea connecting many areas of mathematics.

Chapter 7

Homogeneous Coordinates: A Concept to Better Understand Infinity

As we saw in Chapter 3, we can describe the location of places on the globe using two numbers — the longitude and the latitude. Vienna, the capital of Austria, is located at 48.21° N latitude and 16.37° E longitude. In geometry, this pair of numbers is referred to as the spherical coordinates. The powerful concept of describing the position of a point in a given space — which can be, for example, the Euclidean plane, Euclidean space, or the surface of a sphere — using numbers reverts back to the French mathematician, philosopher and natural scientist René Descartes (1596–1650). He was one of the first to solve geometric problems mathematically by combining geometry and algebra. The rectangular Cartesian coordinate system in the Euclidean plane is named after him, although he may not be the original inventor of this concept. We shall consider this in the following section.

7.1 The Rectangular Cartesian Coordinate System

If we want to describe the position of two points A and B in the Euclidean plane in relation to each other using numbers, the most natural measurement is the distance between them as we show in Figure 7.1.

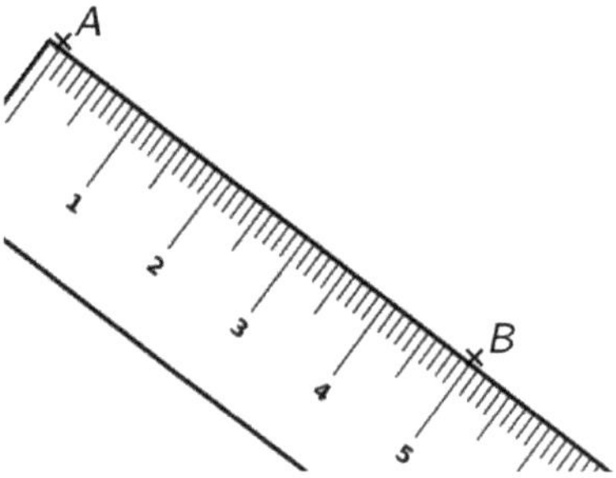

Figure 7.1. Measuring the distance between two points A and B in the Euclidean plane.

By specifying the distance r in Figure 7.1, we have $r = 5$ units. We know that if we place one of the points — say, point A — freely anywhere on the plane, the other point B must lie somewhere on a circle of radius r centered at A as seen in Figure 7.2.

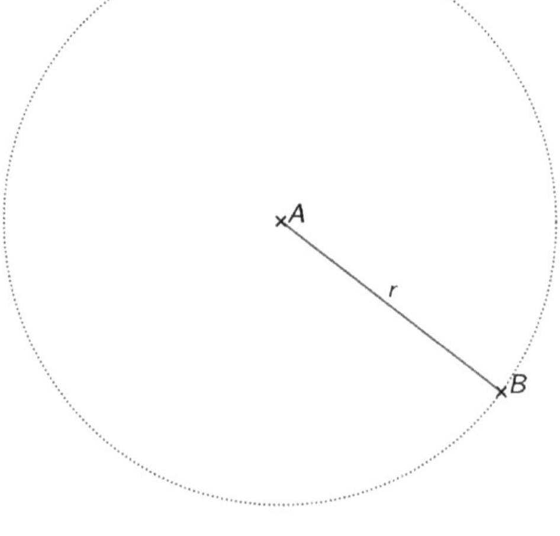

Figure 7.2. Point B lies on a circle with radius r and center A.

So, there are infinitely many possible locations for point B that satisfy the given distance. Now, suppose we add a third point C, and the distances from C to both A and B are also known. We choose an arbitrary position for B on the circle with center at point A, and then determine the possible positions of point C by intersecting two circles: one centered at A with radius $r_A = AC$, and one centered at B with radius $r_B = BC$. The intersection points of these two circles give the possible locations of point C. In general, there are two such points C_1 and C_2, shown in Figure 7.3, corresponding to two possible placements of triangle ΔABC that preserve the given distances.

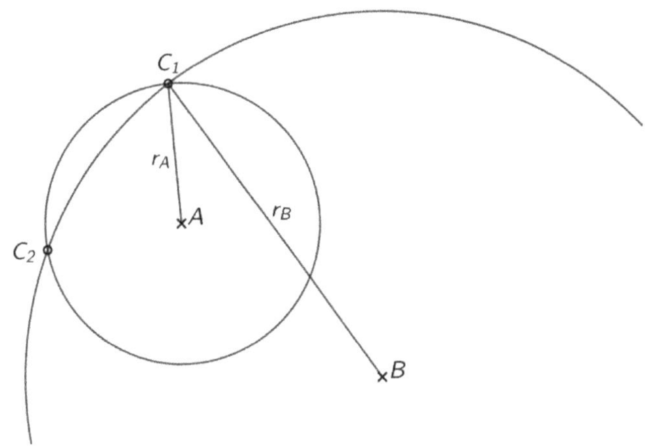

Figure 7.3. Construction of the points by their distances to each other.

Consider Figure 7.4, where two right-angled rays x and y are emanating from a point O. We call them the x-axis and the y-axis. A scale is marked on each of the rays. The two scales start at point O at value 0. The point O is described by the pair of numbers (0,0), as it lies on the first scale on the x-axis (first value) as well as on the second scale on the y-axis (second value) at 0. We call this geometric arrangement a rectangular coordinate system. There are also coordinate systems in which there is no right angle between the two axes, however, we will limit ourselves to this common case, the perpendicular axes, in the following observations.

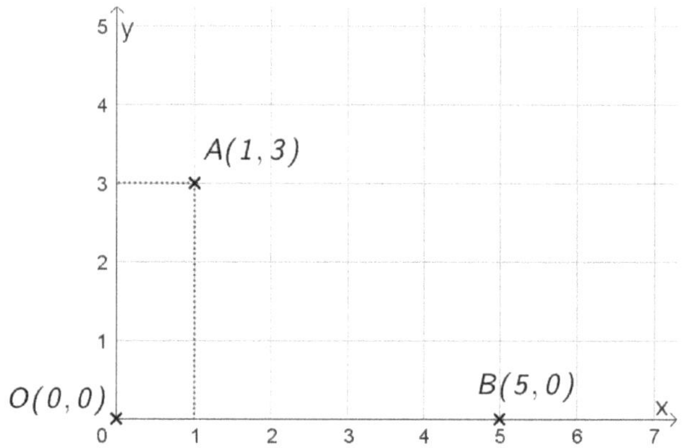

Figure 7.4. The (rectangular) Cartesian coordinate system.

Point A in Figure 7.4 can be described by the pair of numbers (1,3), because we reach point A by moving 1 unit from the origin O in the direction of the x-axis, and then 3 units in the direction of the y-axis. This pair of numbers is called the coordinates of point A. Using the same procedure, we can describe point B as (5,0).

Given any set of points in the Euclidean plane, we can always introduce a coordinate system and assign to each point a pair of numbers that specifies its location. If needed, we can also extend the axes to include negative values, allowing points to have negative coordinates as well, as shown in Figure 7.5.

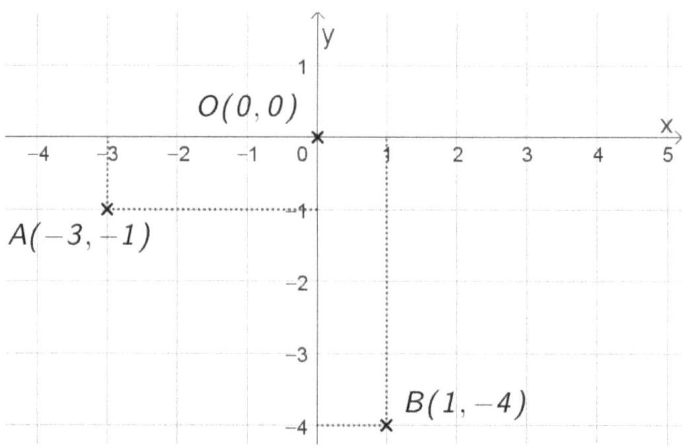

Figure 7.5. Extending the coordinate axes to include negative values.

When we compare Figure 7.4 and Figure 7.5, we see that points A and B have different coordinates, but the relative position between them remains the same. This shows that the choice of coordinate system does not affect how the two points are situated with respect to each other. We can also recognize this by looking at the differences in their coordinates:

In Figure 7.4, point A has coordinates (1,3), and point B has coordinates (5,0). To compare their positions, we subtract the coordinates of A from those of B. The difference in the x-coordinates is $5 - 1 = 4$, and the difference in the y-coordinates is $0 - 3 = -3$.

In Figure 7.5, point A has coordinates $(-3,-1)$, and point B has coordinates $(1,-4)$. Again, subtracting the coordinates of A from those of B, we find: the x-difference is $1-(-3)=4$, and in the y-difference is $-4-(-1)=-3$.

So even though the absolute coordinates differ, the differences between the points — that is, how far and in which direction B lies from A — are exactly the same in both cases.

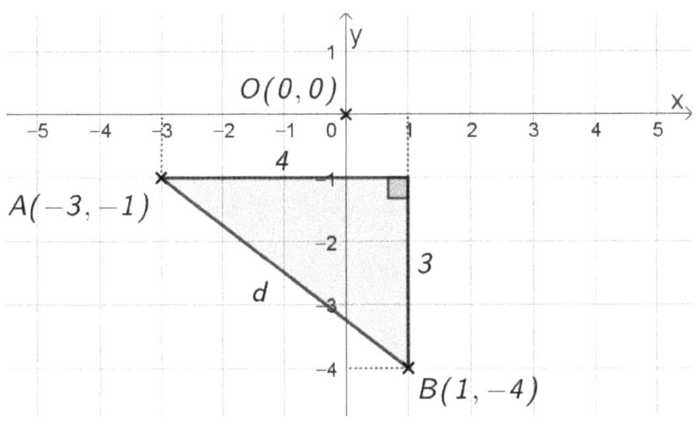

Figure 7.6. Calculating the distance d between A and B.

Now consider Figure 7.6, where we have the same figure as shown in Figure 7.5 augmented by a right triangle. We can use the coordinates of points A and B to calculate the lengths of the legs, the two sides of a right triangle that are adjacent to the right angle, of this triangle. Recall the Pythagorean theorem that enables us to calculate the length of the third side of a right triangle given the lengths of two sides using the formula $a^2 + b^2 = c^2$, where a and b are the lengths of the legs and c is the length of the hypotenuse, which is the longest side of the right triangle and is opposite the right angle. In our example from Figure 7.6, $a = 3$ and $b = 4$, and we can then calculate d as: $d = \sqrt{3^2 + 4^2} = \sqrt{25} = 5$.

We can, therefore, conclude that if we choose any coordinate system for a given set of points in the Euclidean plane, we can calculate all distances between pairs of given points from the coordinates of the points by using the Pythagorean theorem.

7.2 Describing Points and Straight Lines in the Euclidean Plane Using Coordinates

As we saw in the last section, we can describe the position of a point in the Euclidean plane using a pair of numbers if we choose a (rectangular) Cartesian coordinate system. If a point P is described by the pair of numbers (2,4), we know that we can find it in the plane by measuring from the coordinate origin O first 2 units in the direction of the x-axis and then 4 units in the direction of the y-axis, as shown in Figure 7.7.

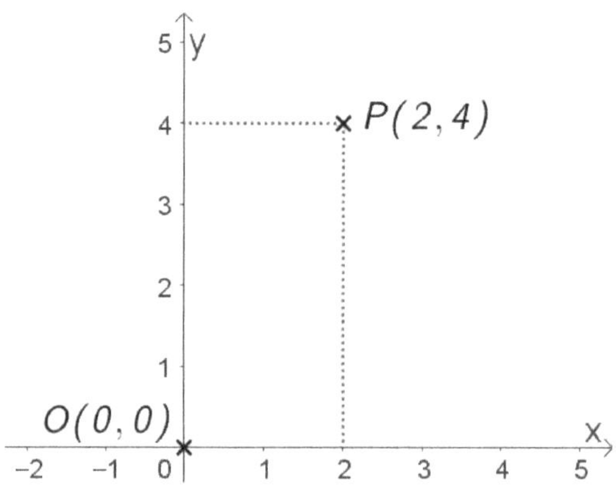

Figure 7.7. Point P is described by the pair of numbers (2,4).

Now that we know how to describe a single point with a pair of numbers (its coordinates), we will apply the concept of description by coordinates to sets of points. Let's start with a line segment s between points A and B as can be seen in Figure 7.8.

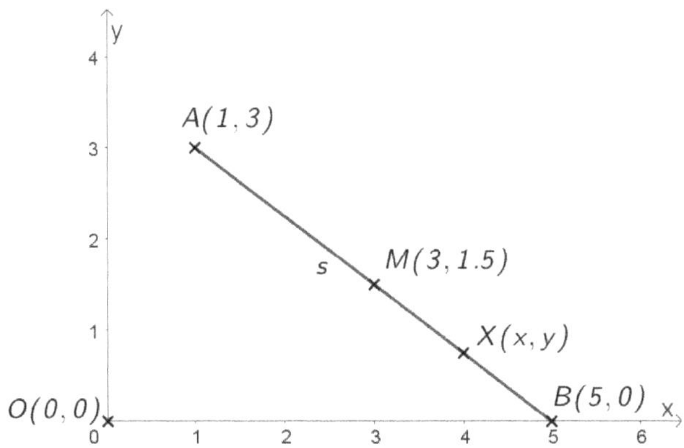

Figure 7.8. Describing points on a line segment using coordinates.

Points *A* and *B* are the endpoints of the segment *s*. As we can easily see, we get the coordinates of the midpoint *M* of the line segment *s* by determining the arithmetic mean of the coordinates of

$$A \text{ and } B: \frac{1+5}{2} = 3 \text{ and } \frac{3+0}{2} = 1.5.$$

Let's try to generalize the concept of how we get the coordinates of the midpoint *M* for other points on *s*. Consider how we calculated the coordinates (x_M, y_M) of *M* from the coordinates $(x_A, y_A) = (1,3)$ and $(x_B, y_B) = (5,0)$ of *A* and *B*:

$$x_M = \frac{1}{2}x_A + \frac{1}{2}x_B = \frac{1}{2} + \frac{5}{2} = 3 \text{ and } y_M = \frac{1}{2}y_A + \frac{1}{2}y_B = \frac{3}{2} + \frac{0}{2} = 1.5.$$

Now, instead of multiplying each of the coordinates by $\frac{1}{2}$, let's try two other fractions that also have a sum of 1, for example, $\frac{3}{4}$ and $\frac{1}{4}$, and calculate the results:

$$x = \frac{3}{4}x_A + \frac{1}{4}x_B = \frac{1}{4} \cdot 1 + \frac{3}{4} \cdot 5 = 4 \text{ and } y = \frac{3}{4}y_A + \frac{1}{4}y_B = \frac{1}{4} \cdot 3 + \frac{1}{4} \cdot 0 = \frac{3}{4}.$$

We actually get another point on *s*: point $X = (x, y) = \left(4, \frac{3}{4}\right)$, as we can see in Figure 7.8. For each value *t* between 0 and 1, the following formulas give us the coordinates of points on the line segment s: $x = t \cdot x_A + (1-t) \cdot x_B$ and $y = t \cdot y_A + (1-t) \cdot y_B$. If we write the coordinates in rows, we can summarize these two formulas in the following way:

$$\begin{pmatrix} x \\ y \end{pmatrix} = t \cdot \begin{pmatrix} x_A \\ y_A \end{pmatrix} + (1-t) \cdot \begin{pmatrix} x_B \\ y_B \end{pmatrix},$$

or more concisely as $X = t \cdot A + (1-t) \cdot B$, where t is chosen between 0 and 1, which we can express it mathematically as follows: $X = t \cdot A + (1-t) \cdot B$, with t ranging between 0 and 1.

What happens if we choose any real value for t that is not between 0 and 1? We can experiment with $t = 2$ and get a point $P(x_P, y_P)$ with the coordinates: $x_P = 2 \cdot 1 + (-1) \cdot 5 = -3$, and $y_P = 2 \cdot 3 + (-1) \cdot 0 = 6$. Figure 7.9 shows that point P lies on the straight line that continues beyond the endpoints of segment s.

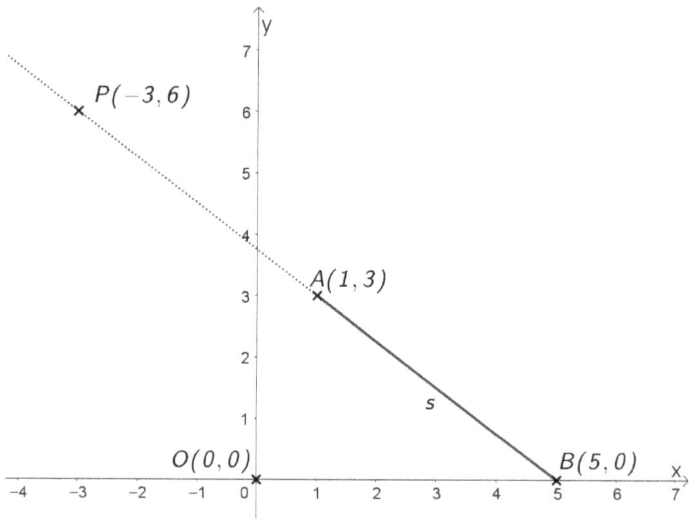

Figure 7.9. Using coordinates to describe a line.

If we choose any real number for t, the following equation gives us points X on the straight line through A and B as $X = t \cdot A + (1-t) \cdot B$, with t ranging between 0 and 1.

We can now also describe connecting straight lines between two points as a point set in the Euclidean plane by means of a coordinate system using numbers (coordinates). For the sake of completeness, it should be mentioned here that an exact proof of our above considerations is beyond the scope of this book.

Since we can now describe the line between two points using coordinates, we now turn to the intersection of two distinct lines. In the Euclidean plane, we have to distinguish between two cases: the two lines can be parallel, in which case we do not get an intersection as in Figure 7.10, or intersecting lines, in which case they have an intersection point, as shown in Figure 7.11.

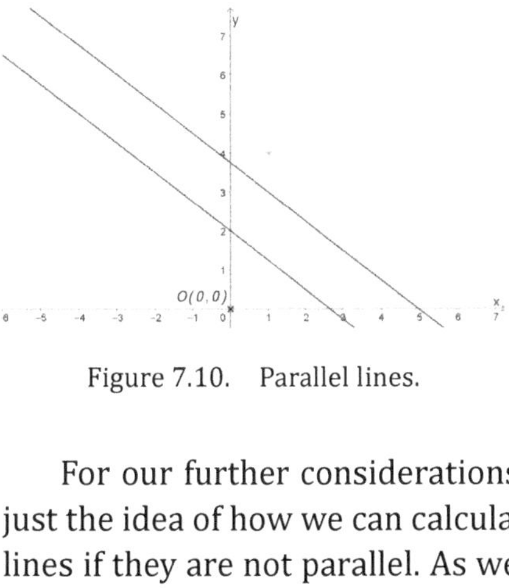

Figure 7.10. Parallel lines.

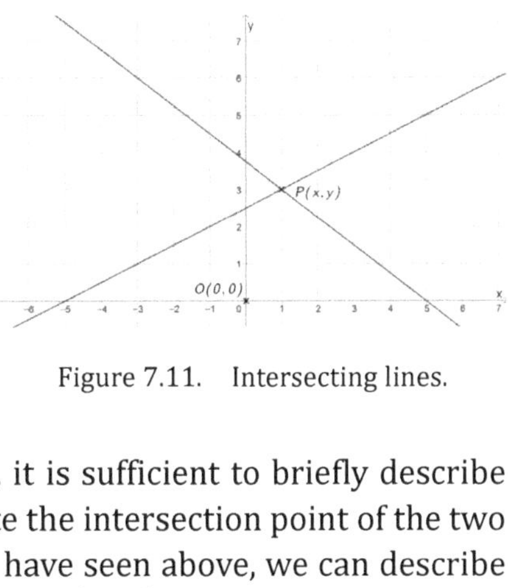

Figure 7.11. Intersecting lines.

For our further considerations, it is sufficient to briefly describe just the idea of how we can calculate the intersection point of the two lines if they are not parallel. As we have seen above, we can describe a straight line g in the form: $X = t \cdot A + (1-t) \cdot B$, where A and B are points on the line and t can be any real number. Another straight line h may pass through the points C and D and can be described as: $X = s \cdot C + (1-s) \cdot D$, and s can be any real number. If we now want to calculate the coordinates of the intersection point $P(x, y)$, we can equate the two above equations:

$$t \cdot A + (1-t) \cdot B = s \cdot C + (1-s) \cdot D.$$

Since we know the coordinates of the points A, B, C and D, we get two equations (one for each coordinate) in the two unknowns s and t. Solving these equations gives us values for s and t, and if we then insert these into the two straight line equations of g and h, we get the same pair of coordinates each time, namely those of the intersection point P.

We can now describe a point, the line connecting two points, and the intersection point of the two (non-parallel) lines with coordinates.

7.3 Describing Circles in the Euclidean Plane Using Coordinates

Another fundamental geometric figure, alongside points and straight lines, is the circle. With a straight edge and a pair of compasses we are able to carry out all sorts of constructions in the Euclidean plane. To perform such constructions using coordinates rather than physical tools, we need a suitable way to represent the set of all points on a circle numerically. In other words, we must find a coordinate-based description of a circle that allows us to calculate with it directly.

For our considerations, let us go back to the elementary geometric definition of a circle as a set of points: "A circle c is the set of all points P that have the same distance r (the radius) from a given point M (the center)." (See Figure 7.12).

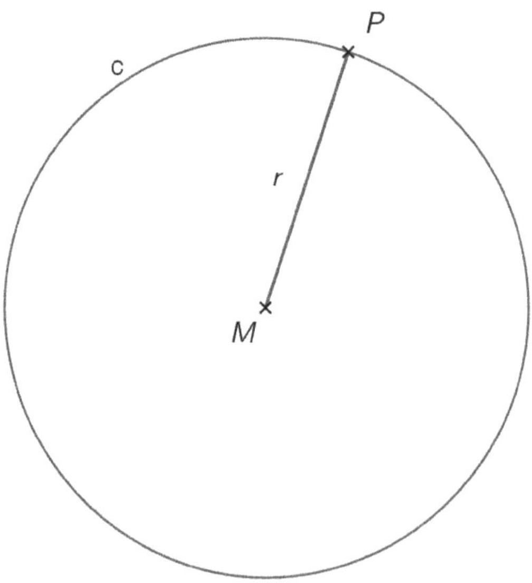

Figure 7.12. A circle c with radius r.

In order to describe a circle as a set of points as simply as possible using coordinates, we choose a special coordinate system, namely, one in which the origin O of the coordinates is at the center M of the circle.

In Figure 7.13 we recognize a right triangle whose legs have the absolute value of the coordinates x and y of each point P on the circle c as their lengths and whose hypotenuse length is the circle radius r. Exactly those points $P(x, y)$ that fulfill the equation $x^2 + y^2 = r^2$ are, therefore, points of the circle c.

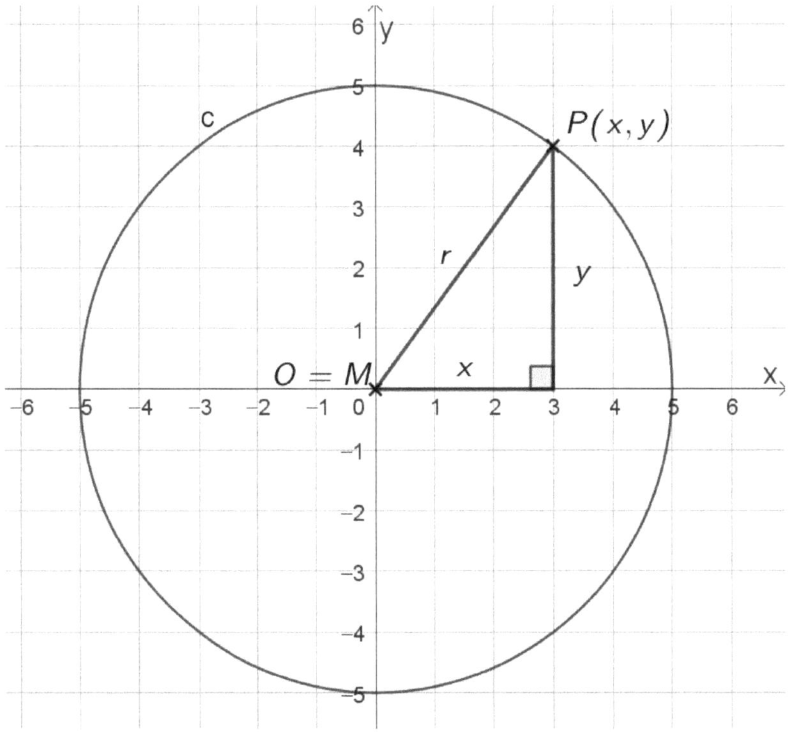

Figure 7.13. A circle c in a special coordinate system.

What happens if we do not choose the coordinate system so cleverly, that is, where the origin O is not also the center $M(x_m, y_m)$ of the circle? In Figure 7.14, we recognize a right triangle again and now realize that our equation differs from the previous one only by a correction due to the difference with the coordinates of the center point. Exactly those points $P(x,y)$ that satisfy the equation

$$(x-x_M)^2+(y-y_M)^2=r^2$$

are, therefore, points of the circle c. The points of circle c in Figure 7.14, for example, satisfy the equation: $(x-2)^2+(y-1)^2=25$, for the plotted point P is obtained by inserting the coordinates: $(5-2)^2+(5-1)^2=25$, and that is correct.

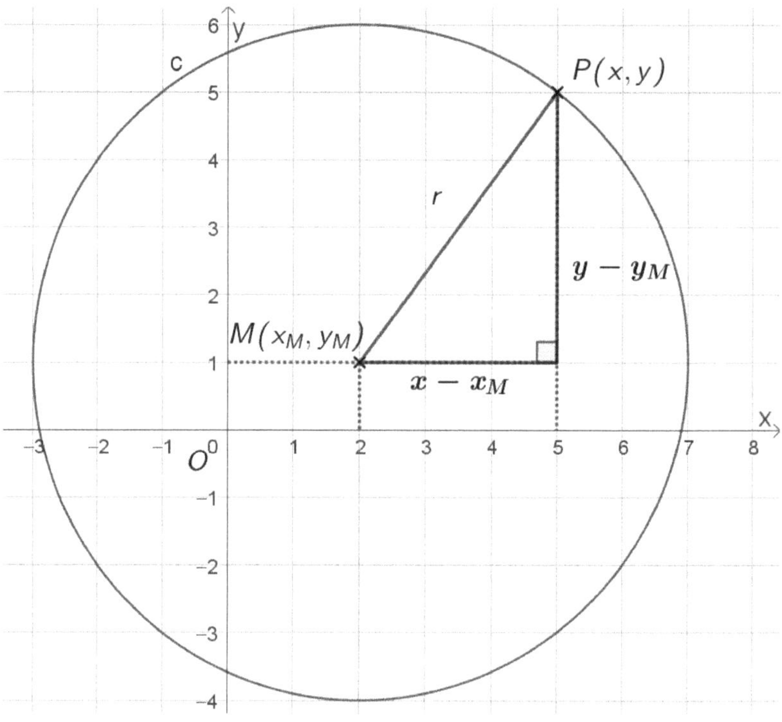

Figure 7.14. A circle c in a coordinate system.

7.4 Points at Infinity and Their Representation Using Homogeneous Coordinates

As we have already seen in Chapter 4, to transform the familiar Euclidean plane into a projective plane, we must supplement it with points at infinity — one for each family of parallel lines. So far, we have seen how Cartesian coordinates allow us to locate points in the

Euclidean plane. But how can we represent points at infinity, which lie beyond the usual coordinate grid?

The basic idea is to introduce a third number — a kind of extra coordinate — that tells us whether we are dealing with an ordinary point in the plane or a point at infinity. Instead of describing a point with two numbers like (x,y), we now use three numbers, written as $(x:y:z)$.[1] This is called a system of homogeneous coordinates.

These coordinates work a bit differently than what we are used to as they do not represent a single point uniquely. For example, the triples $(1:2:3)$ and $(2:4:6)$ describe the same point, because they differ only by a scalar factor. What matters is the ratio between the numbers.

Here is the key idea:

- If the third number z is not zero, the point corresponds to an ordinary location in the plane — just as which are accustomed to.
- But if $z = 0$, then we are dealing with a point at infinity. These special points represent directions, like the "vanishing point" of a family of parallel lines (see Chapter 4.3).

Using this system, we can describe all points — both ordinary and infinite ones — with the same kind of coordinates. That is what makes projective geometry so elegant: it treats finite and infinite points on equal footing.

Let's consider two examples of points given in homogeneous coordinates.

The point $P=(2:3:1)$ is an example of an ordinary point in the Euclidean plane, since the third coordinate $z=1\neq 0$. To convert it to familiar Cartesian coordinates, we divide all components by $z = 1$, giving us

$$\left(\frac{2}{1},\frac{3}{1}\right)=(2,3).$$

This means the point lies 2 units to the right and 3 units up from the origin as shown in Figure 7.15.

---

[1] Strictly speaking, the coordinate triple should be written as a column vector, that is, transposed — but for our purposes, this notation is sufficient.

The point $Q = (1:-2:0)$ is an example of a point at infinity, since its third coordinate is $z = 0$. In projective geometry, such points do not have a fixed location in the plane, but instead represent directions. Specifically, the ratio $1:(-2)$ describes the direction of all lines with slope -2 — that is, lines that go down 2 units for every 1 unit to the right. All such lines are said to "meet" at this point at infinity. These points are known as ideal points and are a fundamental part of projective geometry, allowing parallel lines to intersect in a consistent way. The point at infinity Q is shown in Figure 7.15 using three arrows that indicate parallel directions converging at a common ideal point.

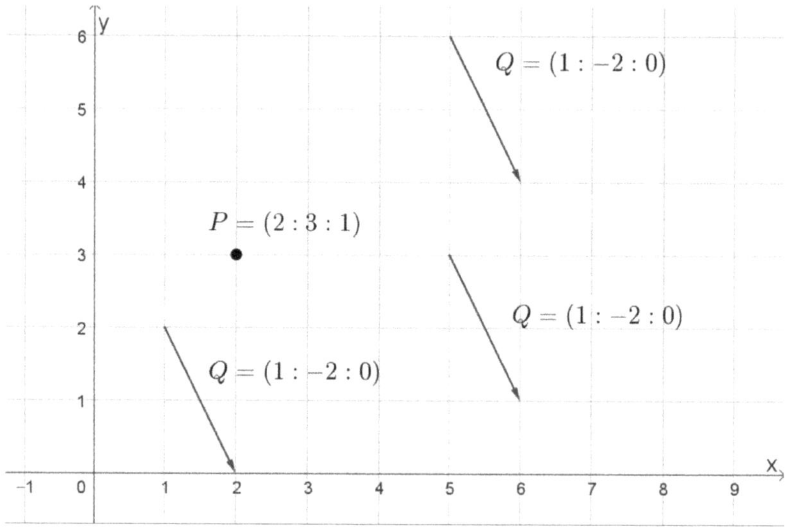

Figure 7.15. Two examples of points given in homogeneous coordinates.

These two examples illustrate how homogeneous coordinates allow us to describe both finite and infinite points in a single, unified framework.

It is a curious twist in projective geometry: the only triple that does not represent a point — namely (0:0:0) — is the one that actually has a name. It is often referred to as the "unpoint," precisely to emphasize that it lies outside the entire geometry. While all other homogeneous coordinate triples describe either ordinary points or points at infinity, this one marks the limit of what the system can express. It is a neat example of how even a non-existent point can play a significant role — if only to remind us of the limits to geometry.

To work comfortably with both ordinary points and points at infinity, it is useful to understand how to convert between Cartesian and homogeneous coordinates. To convert from Cartesian coordinates (x,y) to homogeneous coordinates, we simply add a third coordinate and write $(x{:}y{:}1)$. This embeds the point into the projective plane as an ordinary (finite) point. Conversely, to go from homogeneous coordinates $(x{:}y{:}z)$ back to Cartesian form, we divide the first two coordinates by the third (assuming $z \neq 0$), giving $\left(\frac{x}{z}, \frac{y}{z}\right)$. If $z = 0$, the point lies at infinity and has no Cartesian representation.

By combining Cartesian and homogeneous coordinates, we can now perform calculations in the projective plane, even with points at infinity.

7.5 Describing Lines with Homogeneous Coordinates

We usually describe lines with equations like $y = mx + b$. That works fine for most lines — as long as they are not vertical. A vertical line like $x = 4$ does not have a defined slope, so it does not fit this form.

In projective geometry, however, we have a more general and flexible way to describe lines — one that works for all directions, including vertical lines and even lines "at infinity." In this system, we describe a line using three numbers, called homogeneous coordinates, written as $(a : b : c)$. These define the equation of a line in the form

$$ax + by + cz = 0.$$

Any point $(x : y : z)$ that satisfies this equation lies on the line.

If we take the line $g: y = 2x + 3$, for instance, we can rearrange it into the form $2x + y - 3 = 0$. To convert this equation into homogeneous coordinates, we replace x with $\frac{x}{z}$ and y with $\frac{y}{z}$, then multiply through by z to eliminate the denominators:

$$-2\cdot\frac{x}{z}+\frac{y}{z}-3=0\Rightarrow -2x+y-3z=0.$$

The homogeneous representation of the line is given by the triple $(-2:1:-3)$.

To check, for example, whether the point $P=(1:5:1)$ lies on the line $g=(-2:1:-3)$ we compute:

$$a\cdot x+b\cdot y+c\cdot z=-2\cdot 1+1\cdot 5+(-3)\cdot 1=0.$$

The result of the calculation is zero, which means the point P lies on the line g, as shown in Figure 7.16. If it were not, then the point would not lie on the line.

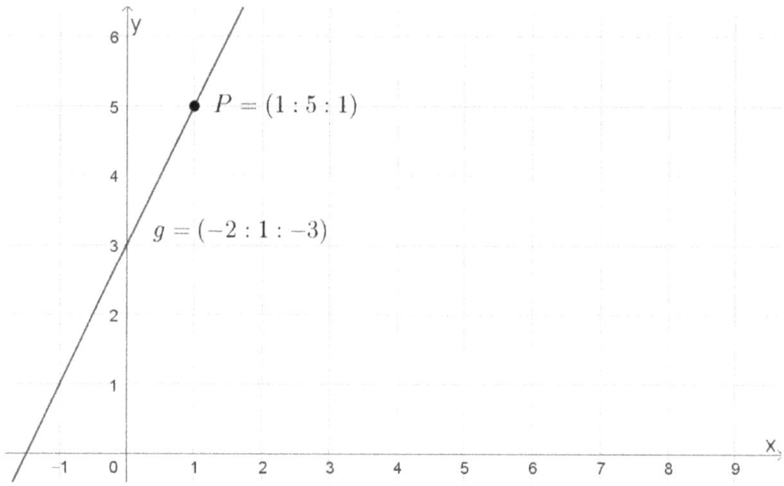

Figure 7.16. A point P lies on the line g.

A vertical line such as $x = 4$, becomes $x - 4 = 0$ in standard form. To convert this into homogeneous coordinates, we replace x with

$\frac{x}{z}$ and y with $\frac{y}{z}$, then multiply through by z to eliminate the denominators:

$$\frac{x}{z} - 4 = 0 \Rightarrow x - 4z = 0.$$

So, the homogeneous representation of the line is: $(1:0:-4)$.

And finally, there is a special line called the line at infinity. It contains all the points at infinity — that is, it captures the idea that parallel lines "meet" somewhere far away. In homogeneous coordinates, it has the simple equation $z = 0$, and is represented by the triple (0:0:1). This line plays a key role in projective geometry, allowing us to treat intersections and parallelism in a consistent and elegant way.

It is important to keep in mind that points and lines are represented differently in homogeneous coordinates. A point lies at infinity if its third coordinate is 0, such as with (x:y:0). The line at infinity, on the other hand, is given by the equation z = 0, which corresponds to the line (0:0:1). This may seem confusing at first — the line at infinity has third coordinate 1, not 0 — but that is because we are now describing a line, not a point.

So, whether a line is tilted, vertical, or infinitely far away — projective geometry lets us describe them all with a single type of equation.

7.6 Describing Circles with Homogeneous Coordinates

As we saw in Chapter 3.3, circles in the Euclidean plane can be described using a Cartesian coordinate system. A circle is defined as the set of all points that are a fixed distance (the radius) from a central point. Algebraically, this leads to the equation:

$$(x - x_M)^2 + (y - y_M)^2 = r^2$$

where center is center $M(x_M, y_M)$, and r is the radius of the circle.

To convert this standard Euclidean circle equation $(x - x_M)^2 + (y - y_M)^2 = r^2$ into homogeneous coordinates, we simply replace x with $\frac{x}{z}$ and replace y with $\frac{y}{z}$, then clear denominators by multiplying through by $z^2 : (x - x_M \cdot z)^2 + (y - y_M \cdot z)^2 = r^2 \cdot z^2$.

This gives us the homogeneous form of the circle:

$$(x - x_M \cdot z)^2 + (y - y_M \cdot z)^2 - r^2 \cdot z^2 = 0.$$

This version directly reflects the geometry of the original circle, and it now lives in the projective plane, including its behavior at infinity.

When we set $z = 1$ in the homogeneous equation of a circle, we return to the familiar Euclidean setting. The equation $(x - x_M \cdot z)^2 + (y - y_M \cdot z)^2 - r^2 \cdot z^2 = 0$ becomes $(x - x_M)^2 + (y - y_M)^2 = r^2$ which is the standard form of a circle with center $M(x_M, y_M)$, and r is the radius of the circle. So, setting $z = 1$ recovers the visible, finite circle in the plane.

But what happens at infinity? To explore this, we set $z = 0$, which corresponds to points "at infinity" in projective geometry. The equation then becomes $x^2 + y^2 = 0$. This equation has no real solutions, since the sum of two squares is never zero unless both are zero — which would give us the invalid point (0:0:0), not allowed in projective geometry. However, if we allow complex numbers — including square roots of negative numbers — then the equation does have solutions: $(1 : i : 0)$ and $(1 : -i : 0)$. These two complex points are called the absolute points, and they lie at infinity. Interestingly, they lie on every circle in the projective plane. That is why they are so important, as they capture a kind of "direction" shared by all circles, and they help unify circle geometry in the projective world. Because the absolute points lie on every circle in the projective plane, they help explain a classical result from algebraic geometry: any two circles intersect in exactly four points, when counted correctly. Depending on the specific case, these four points might all be real and distinct, or include complex number points, or even just one real point counted twice (a so-called double point) — but the total always adds up to four. The two absolute points at infinity are always included in this count. They are "hidden" intersections shared by all circles, and they ensure that circle geometry remains consistent and unified, even when extended into the complex or projective setting.

Even though we are now entering slightly more advanced territory, this book cannot do without the absolute points. They play a key role in connecting projective geometry to classical ideas such as lengths and angles. Through constructions involving these special points at infinity, concepts like harmonic sets and cross-ratios gain new geometric depth — and even notions of distance and angle can be approached without relying on traditional measurement tools. In this way, the absolute points quietly enrich the projective framework and bring it closer to our familiar Euclidean intuition.

In this chapter, we have seen how homogeneous coordinates allow us to perform calculations with points at infinity within the projectively completed plane. By also admitting complex points on the line at infinity, we were able to identify the absolute points, which play a central role in projective geometry — especially in the connection between projective and classical notions like angles and distances.

Just as a side note: in three-dimensional projective space, this role is taken over by the absolute conic, which lies entirely in the plane at infinity and serves a similar unifying purpose for spatial geometry.

Chapter 8

Conic Sections Revisited

Conic sections — ellipses, parabolas, and hyperbolas — arise from the simple yet rich idea of slicing a cone with a plane. This classical geometric construction, rooted in Euclidean space, has fascinated mathematicians since antiquity. By varying the angle and position of the cutting plane, one obtains different types of curves, each with its own intrinsic beauty and mathematical depth. In the first part of this chapter, we will explore how these curves emerge in the Euclidean setting and how their basic structure and relationships can be understood purely through classical geometry.

However, Euclidean geometry captures only one aspect of the story. Projective geometry offers a deeper and more unified perspective, in which many of the distinctions between conic sections begin to dissolve. In this broader framework, conic sections can be generated not only by slicing cones but also purely through projective mappings — transformations that preserve incidence but not distance or angles. Using projectivities, we will see how conics can be defined in ways that are independent of metric notions, and how this reveals unexpected connections between curves that once seemed fundamentally different. The projective approach, thus, both complements and transcends the Euclidean view, offering a powerful language for understanding the true nature of these curves.

8.1 Sections of a Cone

Before we can speak of conic sections, we must first understand what a *cone* is. In geometry, a cone is not just the familiar solid shape with a circular base and a pointed tip. Instead, it refers to an idealized surface formed when a straight line rotates around an axis that it intersects. Imagine a line sweeping around like the hand of a clock — but instead of staying in a plane, it traces out a three-dimensional surface. Because the line crosses the axis, the rotation produces two symmetrical parts: one cone pointing upward, the other downward, both joined at a sharp point A called the *apex*. This creates an infinite, double-sided surface known in mathematics simply as a cone, as shown in Figure 8.1.

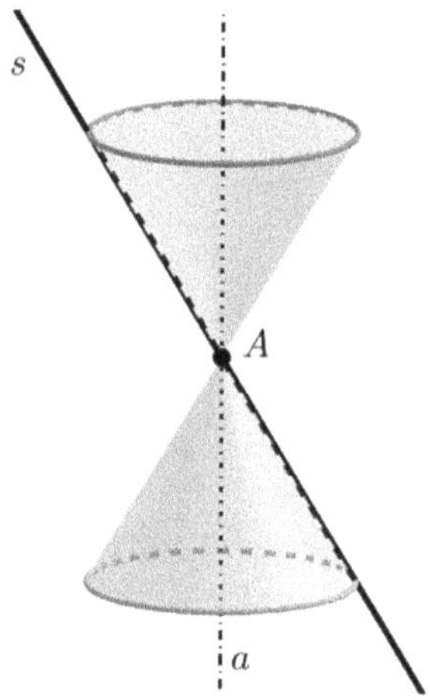

Figure 8.1. Generation of a cone.

When a flat plane cuts through this cone, the intersection forms what we call a *conic section*. The shape depends on the angle and position of the cut: a tilted slice creates an ellipse, a parallel cut yields a parabola, and a steep cut that goes through both halves produces a hyperbola, as shown in Figures 8.2–8.4. For clarity, the cone in these diagrams is shown as truncated — that is, both its tip and base have been sliced off. This makes the curves easier to see, but it is important to note that the flat surfaces where the cone is opened or capped (such as the circular base or the cut-off top) are not considered part of the conic sections. Mathematically, only the curved intersection between the slanted cone surface and the cutting plane defines a conic section.

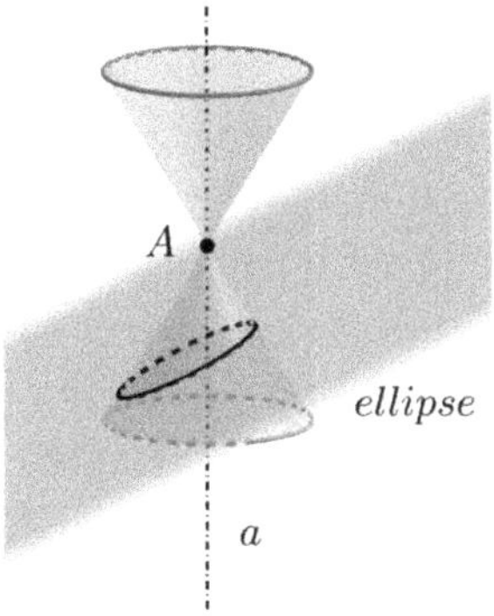

Figure 8.2. Ellipse.

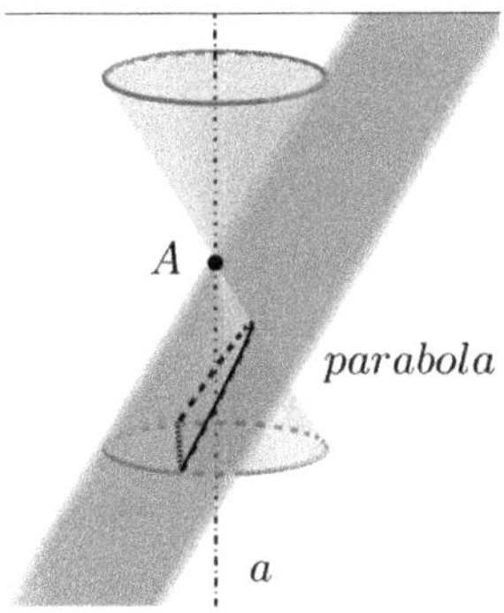

Figure 8.3. Parabola.

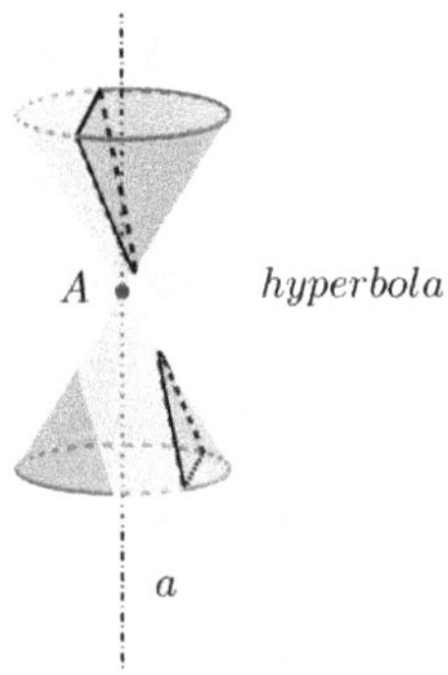

Figure 8.4. Hyperbola.

But these are not the only possibilities. If the plane merely touches the apex, the result can be something quite different: a single point, a straight line, or even two crossing lines — called a degenerate conic. If the cut is perfectly horizontal (perpendicular to the axis of the cone), and does not pass through the apex, we get a special case of an ellipse, that is, a circle. These edge cases may seem unusual, but they help complete the picture and show how all conic sections are part of one continuous geometric family.

8.2 Conic Sections as Plane Curves

Let us now take a closer look at the three classical types of conic sections: the ellipse, the parabola, and the hyperbola. Each of these curves has its own geometric character and its own role in nature and science.

We will begin with the ellipse — a smooth, closed curve that resembles a stretched circle. Among the three, it is the only one that forms a finite, bounded shape. A simple and elegant way to draw an ellipse is the string method, often used by gardeners or architects. To try it yourself, fix two pins in a board — these represent the *foci* of the ellipse, as we show in Figure 8.5. Then tie a loop of string with a fixed length, longer than the distance between the two pins. Place a pencil inside the loop, pull the string taut, and move the pencil while keeping the string taut. The curve traced by the pencil is an ellipse. This construction works because the total length of the string remains constant during the motion. At every position, the string stretches from the pencil to one focus, then to the other — so the sum of the distances to the two foci stays the same. This setup is illustrated in Figure 8.5, where the pins, string, and traced curve are shown clearly.

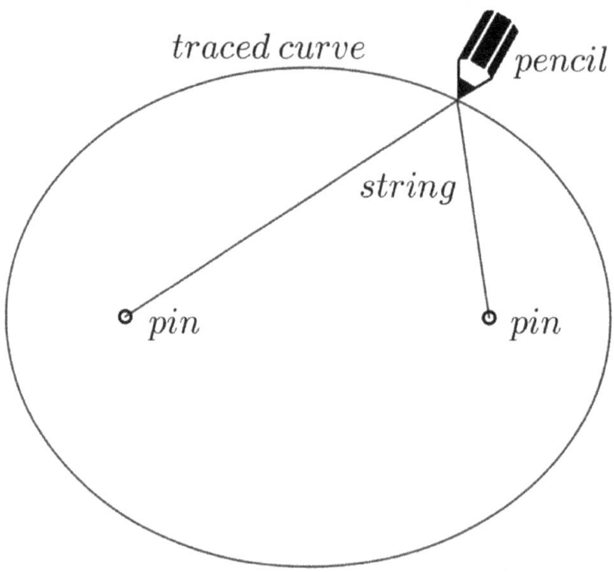

Figure 8.5. The string method for constructing an ellipse.

This leads directly to a mathematical definition: An ellipse *ell* is the set of all points in the plane for which the sum of the distances to two fixed points (the foci) is constant. In symbols: $ell = \{P \in \mathbb{R}^2 : PF_1 + PF_2 = 2a\}$. The symbol $\mathbb{R}^2$ refers to the Euclidean coordinate plane — the set of all points with real-valued coordinates (x, y). In Figure 8.6, F_1 and F_2 are the foci, and $2a$ is the total length of the string.

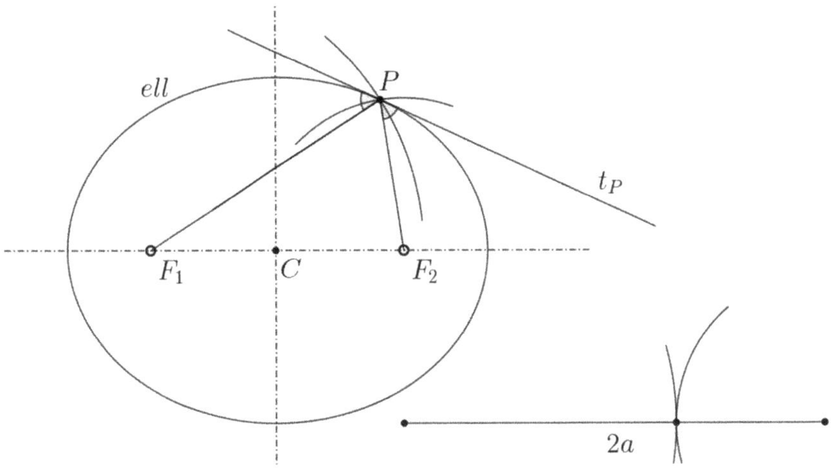

Figure 8.6. Mathematically labeled ellipse.

In Figure 8.6, as we said, we have the foci F_1 and F_2, a point P on the curve, and the corresponding tangent line t_p at that point, constructed geometrically. The point P is constructed using the fixed string length $2a$ in a purely geometric way: the total length is laid out as a straight segment and then divided into two parts. One part is measured from F_1, the other from F_2. With compass circular arcs centered at the foci, the intersection point marks a location P on the ellipse, since it satisfies the condition $PF_1 + PF_2 = 2a$, such points form the ellipse. In this construction, the two symmetry axes of the ellipse also emerge naturally. They are shown in dash-dotted lines: the major axis passes through the two foci, and the minor axis is perpendicular to it at the center C of the ellipse. Since the total string length $2a$ is divided into two segments and each segment can be laid off from its respective focus in two possible directions (left/right or up/down), there are four possible intersection points for a given division — one in each quadrant around the center. These four points lie symmetrically on the ellipse.

The tangent line t_p at point P can now be constructed as the angle bisector between the lines PF_1 and PF_2. Since P lies on the ellipse, the sum $PF_1 + PF_2$ is constant, and small displacements must preserve this total distance — geometrically, this happens precisely when the path "bounces off" the curve at equal angles to both segments. This geometric construction reveals the reflection property of the ellipse: a ray emanating from one focus and reflecting off the ellipse will always pass through the other focus. This property is not only elegant — it also underlies applications in acoustics, optics, and engineering, from whispering galleries to satellite dishes.

The connection between this pointwise definition of the ellipse as a plane curve and its classical construction as a section of a cone is established by a remarkably elegant argument known as Dandelin's sphere construction. This idea was introduced by the Belgian-French mathematician Germinal Pierre Dandelin (1794–1847), who provided one of the most beautiful links between classical geometry and modern definitions.

In Figure 8.7, this relationship is illustrated and explained: two spheres are placed inside the cone so that each is tangent to both the cone and the cutting plane. These are known as the Dandelin spheres. Each sphere touches the cutting plane at exactly one point — let us denote these points as P_1 and P_2. These are the points of tangency, and they lie on the intersection circles of the spheres with the plane.

Now consider any point P on the curve formed by the intersection of the cone and the plane. The key insight is this: the lines from the cone's apex A to point P are tangent to the two spheres. Therefore, the segment from P to the point of tangency P_1 is equal in length to the segment from P to the corresponding focus F_1, and similarly for P_2 and F_2. It follows that:

$$PF_1 + PF_2 = PP_1 + PP_2 = constant.$$

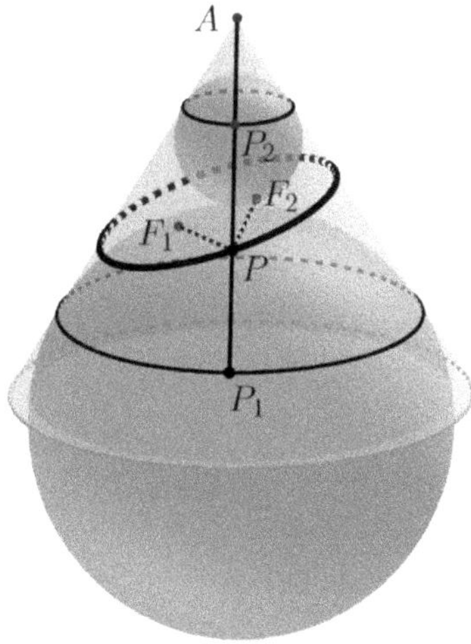

Figure 8.7. Dandelin's sphere construction.

This proves that the intersection curve satisfies the ellipse condition, that is, the sum of the distances to the two foci is constant. Dandelin's construction, thus, elegantly unites the geometric origin of conic sections with their modern pointwise definitions.

Let us now turn to the hyperbola. Just like the ellipse, it can be defined as a set of points in the plane related to two fixed foci — but with a crucial difference: Instead of the sum of the distances to the foci being constant (as in the ellipse), the hyperbola is defined as the set of all points where the absolute value of the difference of the distances to the two foci is constant.

In symbols: $hyp=\{P\in\mathbb{R}^2 : PF_1 - PF_2 = 2a\}$. Here, F_1 and F_2 are the foci, and $2a$ is a positive constant that determines how "open" the hyperbola is. This definition leads to a very different shape: instead of a closed curve like the ellipse, we now get two open branches, each extending to infinity and symmetrical with respect to both a central axis and a center point. As the branches go farther out, they approach two straight lines u and v — called the asymptotes — but never actually touch them. These asymptotes provide a visual frame for the hyperbola and mark the directions in which the curve escapes to infinity. They intersect at the center of the hyperbola and define its large-scale geometry.

There is also a simple geometric construction based on this definition, as shown in Figure 8.8: To find a point P on the hyperbola, one fixes a constant length $2a$ and draws two circles — one with center F_1, the other with center F_2 — such that the difference of the radii equals $2a$. The intersection points of these circles lie on the hyperbola. By varying the pair of radii (while keeping their difference fixed), one obtains different points on the two branches. Through this construction, one again observes the emergence of two symmetry axes: one passing through both foci (the transverse axis) and another perpendicular to it through the center (the conjugate axis). The asymptotes also become visible as limiting directions determined by this symmetrical framework.

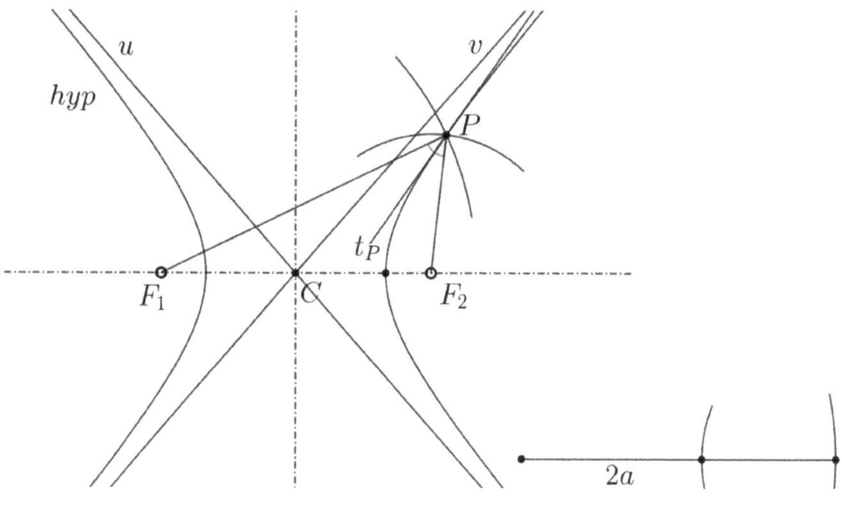

Figure 8.8. Constructing of a hyperbola.

The tangent line t_p at point P can again be constructed as the angle bisector between the lines PF_1 and PF_2.

And just as in the case of the ellipse, this pointwise definition can be connected back to the classical cone construction by an elegant Dandelin's sphere argument. The idea is similar: two spheres are placed inside the cone so that each is tangent to both the cone and the slicing plane, as shown in Figure 8.9. However, in the case of the hyperbola, the two spheres lie in opposite halves of the double cone — one above and one below the cutting plane. Their respective points of tangency with the plane become the foci F_1 and F_2. From the geometry, it then follows that the difference of the distances from any point on the intersection curve to the two tangency points remains constant — thus, confirming the definition of the hyperbola.

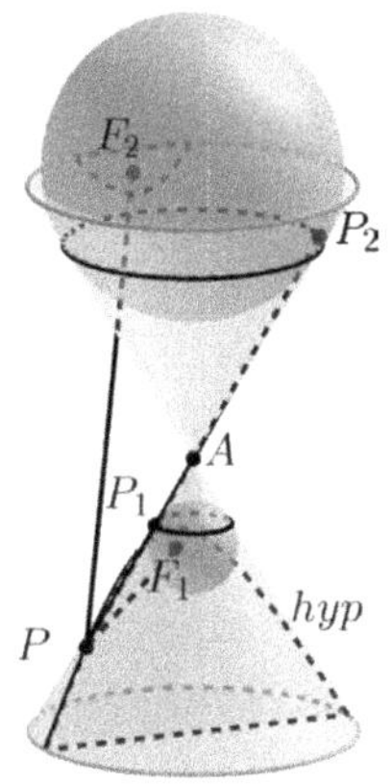

Figure 8.9. Dandelin's construction for the hyperbola.

Let us now turn to the parabola, the third classical conic section. Just like the ellipse and the hyperbola, the parabola can be defined as a set of points in the plane relative to a fixed geometric configuration. In this case, the definition involves one fixed point and one fixed line, which are the focus F and the directrix d. A parabola is the set of all points in the plane that are equidistant from the focus and the directrix: $par = \{P \in \mathbb{R}^2 : PF = Pd\}$. This definition produces a smooth, open curve with a single branch, symmetric with respect to a vertical axis that is perpendicular to the directrix and passes through the vertex of the parabola. The focus lies on this axis, at the same distance from the vertex as the vertex lies from the directrix. To construct a point P on the parabola geometrically, one can proceed as follows: choose a distance a, draw a line parallel to the directrix at distance a, and draw a circle centered at the focus F with radius a. The intersection point P of the circle and the parallel line satisfies the equidistance condition, that is its distance to F and to the directrix is exactly a. Varying a traces out the full parabola. From this setup, the tangent line at a point P on the parabola can be constructed as the angle bisector between the line segment from P to the focus and the perpendicular from P to the directrix. This not only defines the tangent geometrically but also reveals the reflection property of the parabola: Any incoming ray that is parallel to the axis of symmetry reflects off the curve and passes through the focus — and vice versa.

This property explains the use of parabolic mirrors and reflectors in optics, satellite dishes, and car headlights. In a headlight, for example, the bulb is placed at the focus of a parabolic reflector, so that the light rays are reflected outward in parallel beams — maximizing forward illumination, as shown in Figure 8.10. Conversely, in a satellite dish, incoming parallel signals (e.g. from a distant satellite) strike the parabolic surface and are reflected inward, all meeting precisely at the focus, where the receiver is placed, as shown in Figure 8.11. This efficient concentration of energy relies on the same geometric property — and it also holds in three dimensions: when a parabola is rotated around its axis, it generates a paraboloid of revolution, and the focus becomes a true focal point for all incident rays.

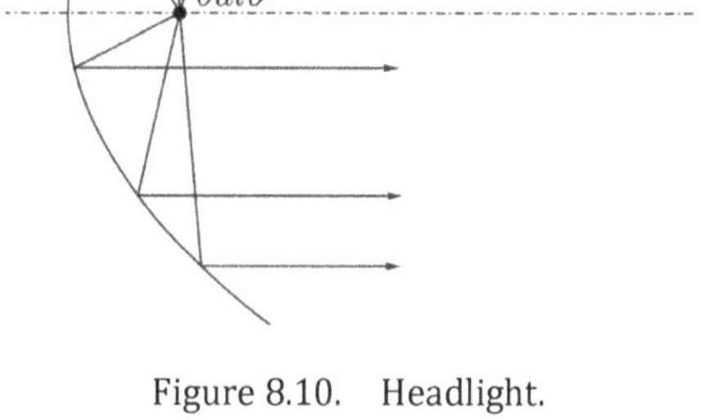

Figure 8.10. Headlight.

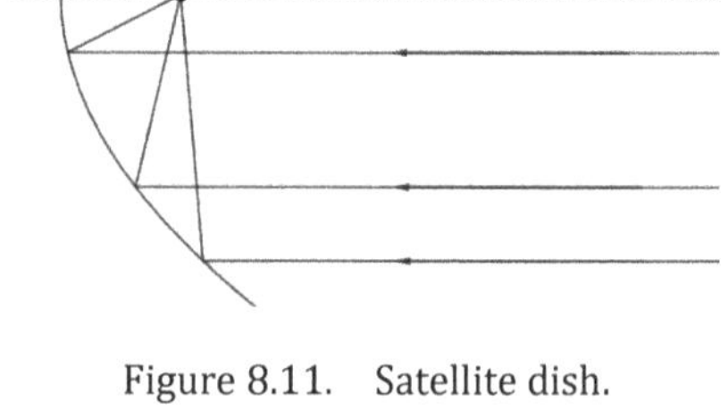

Figure 8.11. Satellite dish.

Similar rotational surfaces arise from the ellipse and hyperbola: rotating an ellipse around its major axis yields an ellipsoid, where light or signals emitted from one focus reflect to the other. This principle underlies the design of elliptical domes and acoustic chambers. Rotating a hyperbola gives a hyperboloid of revolution, which has optical and structural uses as well — for instance, in radio telescopes and cooling towers — where its reflective or load-bearing properties are exploited.

Other applications of planar conics include whispering galleries, elliptical billiards, and hyperbolic navigation systems, all of which rely on the specific geometric definitions and reflection properties of conic sections.

In Figure 8.12, this construction is illustrated: a point P on the parabola is constructed by enforcing the equidistance condition, and the tangent t_p at P is shown as the bisector of the angle between the line to the focus and the perpendicular to the directrix.

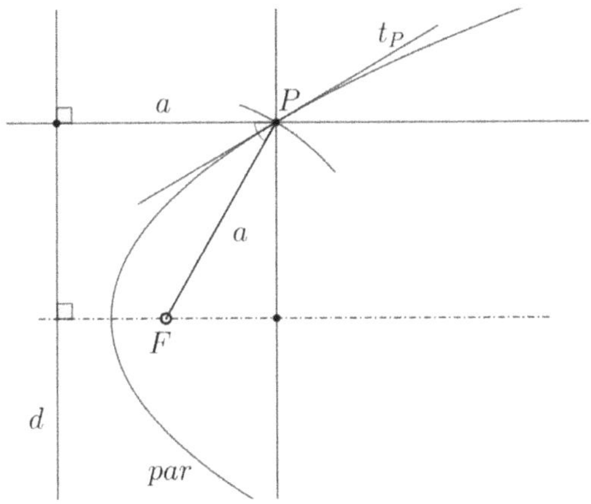

Figure 8.12. Constructing a parabola.

Just as the ellipse and the hyperbola, the parabola also arises as a section of a cone — specifically, when a plane intersects the cone parallel to one of its sides. For the parabola, the connection to the focus–directrix definition requires a bit more geometric reasoning.

In Figure 8.13, we illustrate the key construction. Let P be an arbitrary point on the intersection curve between the cone and the plane. The segments PF and PA, both tangent to the Dandelin sphere, are equal in length. The generating line, which connects the apex A with the point P, lies in a plane that intersects the cutting plane along a line perpendicular to the directrix d. All such lines — generated by varying P — are parallel to each other. Using the intercept theorem on the lines AP and BC, and the parallel segments BP and AC, we conclude that $PA = PB$. Since also $PF = PA$, it follows that $PF = PB$, so the distance from P to the focus F equals its perpendicular distance to the directrix d. This confirms the parabola's defining property.

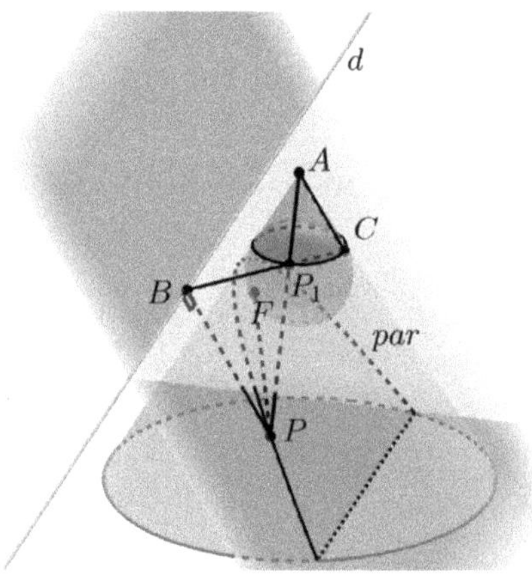

Figure 8.13. Dandelin's method applied to the parabola.

8.3 Conic Sections in Projective Geometry

In projective geometry, conic sections can be studied from two fundamentally different perspectives. One is analytic, using homogeneous coordinates and equations, and the other is axiomatic, based purely on geometric principles such as incidence and cross-ratio.

In this chapter, we adopt the axiomatic approach. Rather than reasoning in terms of lengths or classical constructions — as we did in the previous section within Euclidean geometry — we now take on a completely different perspective. Projective geometry removes the distinction between "finite" and "infinite" and treats all points and lines on equal footing. Concepts like parallel lines or asymptotes are no longer exceptional cases but are naturally incorporated through the introduction of points at infinity and the line at infinity. This shift changes the nature of conic sections profoundly. The classical differences between ellipses, parabolas, and hyperbolas dissolve in the projective plane, where they are all manifestations of a single, unified geometric object — a projective conic.

In Chapter 4, we explored basic concepts of perspectivities and projectivities — special types of geometric mappings between points on lines or rays. Now we return to these ideas from a new angle: in projective geometry, a conic section can be defined purely in terms of such mappings. Specifically, a conic can be described as the set of intersection points of corresponding rays in a projectivity between two pencils of lines (that is, two bundles of rays emanating from different points). At first glance, this definition may sound even more abstract than slicing a cone — but in Chapter 4 we also saw an important fact: any projectivity can be built from at most three perspectivities. In Figure 8.12, we see such a construction in action. A first perspectivity maps the initial pencil $A(x)$ onto a new pencil $C(x')$, with the mapping carried out via an intermediate line $b(X)$; a second perspectivity then maps $C(x')$ onto the final pencil $E(x'')$, again via a line $d(X')$. The resulting correspondence between the initial pencil $A(x)$ and the final pencil $E(x'')$ is a projectivity $A(x) \barwedge E(x'')$, composed of two perspectivities:

$$A(x) \overset{b}{\underset{\wedge}{=}} C(x') \overset{d}{\underset{\wedge}{=}} E(x'').$$

The intersection points of corresponding rays from $A(x)$ and $E(x'')$ form the point set of a conic section. In Figure 8.14, we see a hyperbola generated in this way, where the intersection points of corresponding rays from two projectively related pencils trace out the two branches of the conic. If we now slightly change the position of the point E, the configuration — and, therefore, the conic — changes, where the intersection points no longer form a hyperbola, but an ellipse instead. This can be seen in Figure 8.15.

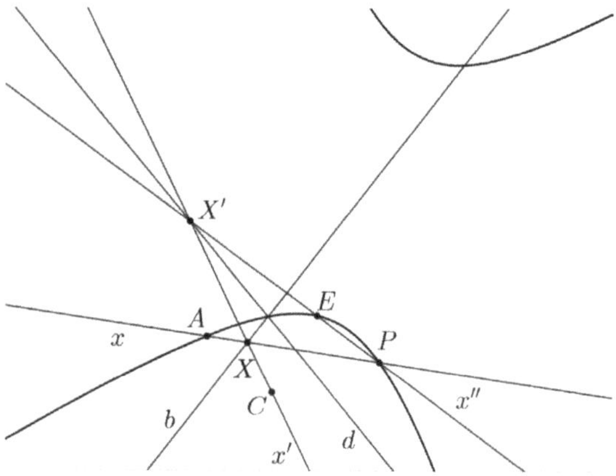

Figure 8.14. Projective construction of a conic section from two pencils.

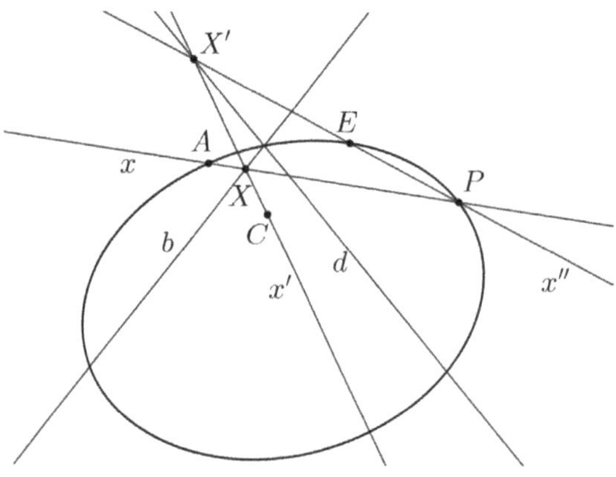

Figure 8.15. Ellipse arising after moving the point E.

According to the above definition — as the set of intersection points of corresponding rays in a projectivity between two pencils — the different types of conic sections (ellipse, hyperbola, and parabola) cannot be distinguished from one another. All of them arise from the same kind of projective construction and are, in this sense, geometrically equivalent.

However, in the projectively completed plane, we can distinguish between conics by looking at the number of intersection points with the line at infinity. This provides a natural classification: ellipses, hyperbolas, and parabolas differ in how they intersect (or touch) the line at infinity.

We will also see that this explains why the parabola did not appear in our earlier constructions based on projectivities between pencils. In those examples, we always obtained conics with two distinct points at infinity — either real (hyperbola) or complex conjugate (ellipse).

The parabola arises as a limiting case, where these two points coincide, and the line at infinity becomes tangent to the conic. It is, therefore, very unlikely to encounter a parabola by accident in a generic projective construction. Unless specific conditions are imposed, the resulting conic will almost always be an ellipse or a hyperbola. This distinction is illustrated schematically in Figures 8.16–8.18, which show how each type of conic intersects the line at infinity.

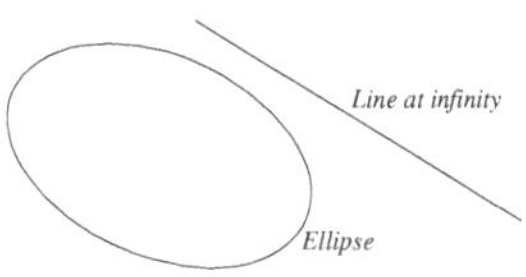

Figure 8.16. Ellipse.

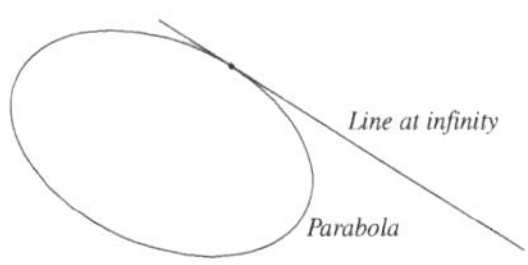

Figure 8.17. Parabola.

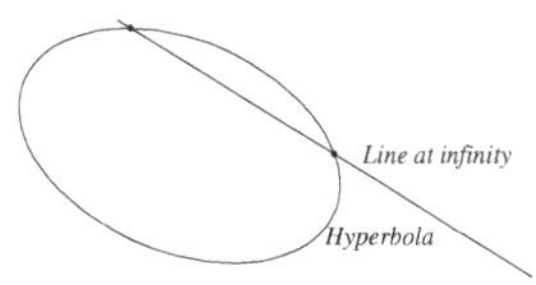

Figure 8.18. Hyperbola.

8.4 Polarity, Cross-Ratios, and Harmonic Configurations on Conics

In projective geometry, conic sections are more than just curved figures — they are rich geometric objects that define deep relationships between points and lines. One of the most fundamental of these is the concept of polarity, which assigns to every point (not on the conic) a uniquely determined line called its polar, and vice versa. This establishes a natural duality between points and lines with respect to the conic.

Closely tied to this is the notion of the cross-ratio (or double-ratio), a projectively invariant quantity defined for four collinear points. When restricted to points on a conic, the cross-ratio allows us to express important geometric configurations — most notably, the concept of harmonic division, where four points form a special relation preserved under projectivities.

We have already encountered both polarity and harmonic configurations in Chapter 6, where we considered them in the special case of the circle as a Euclidean conic. In the current chapter, we now extend these ideas to general conic sections in the projective plane. This broader perspective reveals the deep structural unity of all conics — ellipse, hyperbola, and parabola — as well as their dual relations to lines and points.

To construct the polar line of a point P with respect to a conic c, one proceeds as follows: choose three pairs of points (A, B), (C, D) and (E, F) on the conic such that the chords AB, CD and EF all pass through P. These six points define a hexagon inscribed in the conic. Now consider the three intersection points of the pairs of opposite sides: $Q_1 = AD \cap BC$, $Q_2 = AF \cap BE$ and $Q_3 = BD \cap EF$. Notice that Q_1, Q_2 and Q_3 lie on a single straight line. This line is precisely the polar p of P with respect to the conic c. In this way, the polar line is obtained from the geometry of six points on the conic whose connecting chords intersect in P. The fact that the three constructed points lie on a line is guaranteed by Pascal and reflects the projective duality between points and lines induced by the conic. This can be seen in Figure 8.19.

In addition, if two tangent lines from P to the conic exist, then their points of contact T_1 and T_2 — that is, the points where the tangents touch the conic — also lie on the polar of P. These two trace points are special cases of intersection points of secants and again confirm that the polar line captures the geometric relationship between P and the conic. Together, these constructions illustrate the deep projective duality between points and lines encoded by the notion of polarity.

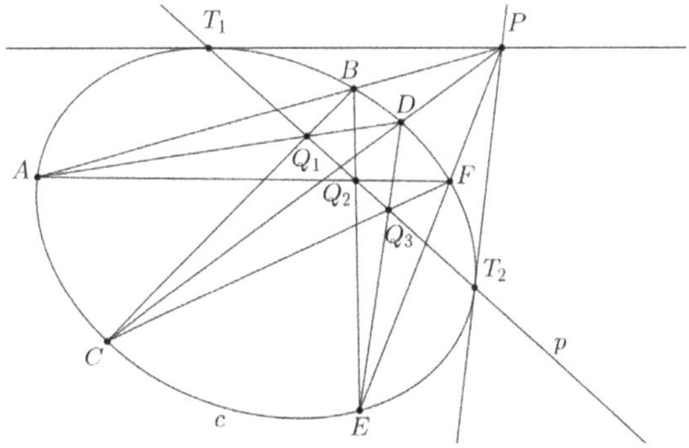

Figure 8.19. Construction of the polar line p of a point P.

A remarkable property of conics in projective geometry is the harmonic relation between a point and its polar as shown in Figure 8.20. Let P be a point outside a conic c, and let p be its polar line. If a secant through P intersects the conic at two points A and B, then the intersection point $X = p \cap AB$ lies on the same line — and the four points A, B, P and X are in harmonic position. In other words, the cross-ratio satisfies $(P, X; A, B) = -1$. This relationship can also be visualized geometrically, as we saw in Chapter 6. In a complete quadrilateral, shown with dashed lines in Figure 8.18, the diagonals intersect in such a way that the harmonic conjugate appears at the intersection point X of the polar and the secant through A and B.

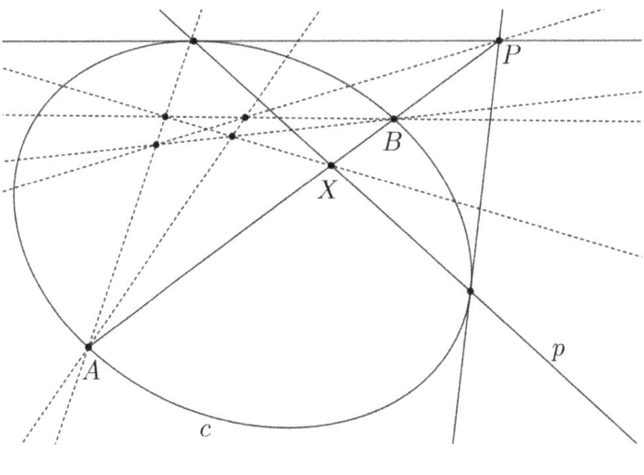

Figure 8.20. Harmonic relation between a point and its polar.

This means that X is the harmonic conjugate of P with respect to A and B, and it balances P in the precise projective sense defined by the conic.

This elegant result goes back to the Swiss mathematician Jakob Steiner (1796–1863), one of the leading figures of 19th-century synthetic geometry. The harmonic property of polar constructions arises from the involutive nature of polarity with respect to a conic and reflects the deep interplay between incidence and duality in projective geometry.

Polarity is an example of an involution — a transformation that undoes itself when applied twice. That is, the polar of the pole of a line is the line itself, and the pole of the polar of a point is the point again. This involutive behavior reveals a fundamental symmetry between points and lines defined by the conic.

Although similar ideas were also explored by Jean-Victor Poncelet (1788–1867) in France, the version involving harmonic division is most directly attributed to Steiner's work on projective conics and line systems.

8.5 Further Perspectives on Conic Sections

In this Section, we will briefly touch on several deeper aspects of conic sections that lie beyond the main scope of this book. These topics offer powerful and elegant perspectives — especially within the framework of projective geometry — but we will mention them only in passing, without detailed proofs or formal development. A full treatment would go beyond our intended level and purpose.

One important question concerns how a conic can be uniquely determined. From a projective point of view, five geometric elements — such as points or lines, possibly including complex or ideal (infinite) ones — are sufficient to specify a conic uniquely, provided they are in general position. This leads to a variety of representation methods for conics, depending on which types of elements are used.

Closely related is the rich system of pole-polar relationships, where every conic defines a dual correspondence between points and lines, as we have already seen. But in more advanced settings, these relationships can also be used to define and study geometric configurations, transformations, and symmetry structures — particularly through the language of involutions and autoprojectivities. A striking example is the notion of conjugate diameters in conics, which form an involutive pairing: each diameter corresponds to another such that their directions are related in a symmetric way. This concept has both geometric and algebraic interpretations, especially in affine and metric settings.

Finally, one can make sophisticated use of points at infinity and complex elements in geometric constructions. By cleverly using them as fixed elements of involutions or autoprojective transformations, one can design elegant synthetic constructions — avoiding coordinate calculations while still accessing the full generality of conics in the projective plane.

All of these ideas open rich avenues for further exploration — and for readers who wish to go deeper, they form the foundation of more advanced treatments of conic sections, particularly in higher-level projective and algebraic geometry.

Index

A

absolute points, 132–133
apex (of a cone), 136–137, 140, 146
axioms of projective geometry
  and duality, 61–62
  and Fano axiom, 67, 73–75
  incidence axioms, 56–57
  minimal model, 56–60
axis of perspectivity, 68
autoprojectivity
  definition, 65
  fixed points, 82–83, 87–88
  in problem solving, 82–83, 90

B

Beltrami, Eugenio, 38
Bolyai, János, 38
Brahmagupta's formula, 21–25
  for cyclic quadrilateral, 22–25
  and Heron's formula, 22
Brianchon, Charles Julien, 76
Brianchon's theorem, 76

C

Cartesian coordinate system, 115–120
  circle equation, 124–126
  distance formula, 119–120
  intersection of lines, 123–124
  parametric form of line, 121–123
center of perspectivity, 63, 68
Ceva, Giovanni, 28
Ceva's theorem
  converse, 33–34
  proofs, 29–34
  statement, 28–29
circle
  and absolute points, 132–133
  angle theorems (inscribed, secant, tangent, chord), 9–10
  in homogeneous coordinates, 131–133
  power of a point, 110–111
  secant-tangent theorem, 110–111

collinearity, 95
complete quadrangle, 73–75, 100–103
cone
  Dandelin spheres, 140–143, 146
  generation, 136
conic sections
  classification by line at infinity, 149
  degenerate conics, 137
  ellipse, 137–141
  and involution, 152
  harmonic relation (point and polar), 151–152
  hyperbola, 141–143
  parabola, 143–146
  polarity with respect to a conic, 149–152
  projective definition via projectivity of pencils, 147–149
  from slicing a cone, 135–137
cross-ratio
  on conics, 149–152
  definition (oriented distances), 97–98
  harmonic conjugate construction, 102–103
  harmonic quadruple, 100–102
  and midpoint construction, 104–106
  in polar geometry, 110–112
  invariance under perspectivity/projectivity, 98–99
  permutations of four points, 99
Coxeter, H. S. M., 56, 79

D
Dandelin, Germinal Pierre, 140
Dandelin spheres, 140–143, 146
Desargues, Girard, 55, 68
Desargues' theorem, 67–68
  in Moulton plane, 69–70
Directrix (parabola), 143–146
Duality (in projective geometry)
  Brianchon as dual of Pascal, 76
  dual of Fano axiom, 74
  dual of Pappus's theorem, 72
  dual perspectivity, 64
  principle, 55–56, 61–62

E
elation, 66
ellipse
  Dandelin spheres, 140–141
  definition (sum of distances), 139–140
  foci, 139–141
  major/minor axes, 139
  in projective geometry, 148–149
  reflection property, 140
  string construction, 138–139
  tangent construction, 140
elliptic geometry, 38–47
  angle sum of triangle, 40–42
  great circles, 39–43
  parallels, 44–45
  on a sphere, 39–47
equiangular point, 17–18
Escher, Maurits Cornelis, 5, 48, 51
Euclid, 35
Euclid's fifth postulate, 35–38
  modifications, 38
  Playfair's axiom, 37
  triangle angle sum, 37

F
Fano axiom, 67, 73–75
 Fano plane, 74–75
foci (conic sections)
 ellipse, 139–141
 hyperbola, 141–143
 parabola, 143–146

G
Gauss, Carl Friedrich, 38
geodesics, 48–49
great circle, 39–43

H
harmonic conjugate, 102–103
harmonic quadruple
 construction via complete quadrangle, 100–102
 definition, 100
 of midpoint, 104–106
 in polarity, 110–112, 151–152
Heron's formula, 18–21
Hilbert, David, 57
homogeneous coordinates
 absolute points, 132–133
 circle equation, 131–133
 conversion to/from Cartesian, 127–129
 definition, 126–127
 line representation, 129–131
 line at infinity, 131
 points at infinity, 127–128
homology, 66
hyperbola
 branches, asymptotes, 141–142
 Dandelin spheres, 142–143
 definition, 141–142
 foci, 141–143
 in projective geometry, 147–149
 tangent construction, 142
hyperbolic geometry
 applications, 52–53
 geodesics, 48–51
 parallels, 51–52
 Poincaré disk model, 48–53

I
incidence relation, 56–57
infinity
 line at infinity, 62, 131, 149
 points at infinity (projective), 62, 111–112, 126–129
inversion (circle inversion), 50, 108–109
involution
 on conics, 152
 definition, 109
 in polarity, 109–110, 152

L
latitude and longitude, 39, 115
law of cosines, 10–11
law of sines, 10–11
line at infinity, 62, 131, 149
Lobachevsky, Nikolai Ivanovich, 38

M
midpoint
 via harmonic conjugate, 104–106
 in perspective image, 105–106
minimal model, 56–60
Moulton, Forest Ray, 69
Moulton plane, 69–71, 76

P

Pappus of Alexandria, 71
Pappus's theorem, 67, 71–72
parabola
  applications, 145
  construction, 143–144
  Dandelin sphere, 146
  definition, 143–144
  as limiting case in projective geometry, 149
  reflection property, 144
  tangent construction, 144
parallel lines
  in Euclidean plane, 44–45
  in hyperbolic plane, 51–52
  in projective geometry, 62
  on sphere, 44–45
Pascal, Blaise, 55, 75
Pascal's theorem, 67, 75–76, 86, 150
perspectivity
  definition, 63
  dual perspectivity, 64
Playfair, John, 37
Playfair's axiom, 37
Plücker, Julius, 55
Poincaré, Henri, 38
Poincaré disk model, 48–53
points at infinity
  polar of a point at infinity, 111–112
  in projective completion, 62, 111–112
  representation in homogeneous coordinates, 127–128
polar (line)/polarity
  construction via tangents, 107–108
  and harmonic quadruple, 110–112
  and inverse point, 108–109
  with respect to a circle, 107–112
  with respect to a conic, 149–152
pole, 107, 152
Poncelet, Jean-Victor, 55, 152
power of a point, 110–111
projection (perspective), 3
projective completion of Euclidean plane, 62
projective plane
  axioms, 56–57
  Desarguesian/non-Desarguesian, 68–71
  finite projective planes, 60
  minimal model, 57–60
  order, 60
  Pappian/non-Pappian, 72
projectivity
  composition of perspectivities, 65
  and conics, 147–149
  definition, 65
Ptolemy (Claudius Ptolemaeus), 25
Ptolemy's theorem
  converse, 28
  proofs, 26–28
  statement, 25
Pythagorean theorem, 8, 119

Q

quadrilateral
  Brahmagupta's formula, 21–25
  cyclic, 22–28
  minimum distance point, 17
  Ptolemy's theorem, 25–28

R

reflection property
    ellipse, 140
    parabola, 144
Riemann, Bernhard, 38
right triangle properties, 8

S

Seidenberg, Abraham, 56, 79, 86
similarity of triangles, 8
spherical geometry, 39–47
Steiner, Jakob, 55–56, 88, 152
Steiner circle, 2, 83–89

T

tangents, 9–10
triangle
    Ceva's theorem, 28–34
    congruence conditions, 7
    exterior angle bisector theorem, 13–14
    Heron's formula, 18–21
    interior angle bisector theorem, 12–13
    minimum distance point (Fermat point), 17–18
    right triangle properties, 8
    similarity conditions, 8
    on sphere, angle sum, 40–42
    sum of distances in equilateral triangle, 14–16
trigonometry identities, 11

V

Vanishing point, 95, 105
von Staudt, Karl Georg Christian, 55

Z

Zero vector (unpoint), 128–129

www.ingramcontent.com/pod-product-compliance
Lightning Source LLC
LaVergne TN
LVHW051055210926
848486LV00002B/114